Developing Skills in

ALGEBRA ONE

Harold Taylor ■ Loretta Taylor

Book D

DALE SEYMOUR PUBLICATIONS

Cover Design: Michael Rogondino
Technical Illustrations: Pat Rogondino

Order Number DS01444
ISBN 0-86651-224-1

**DALE
SEYMOUR
PUBLICATIONS**
P.O. BOX 10888
PALO ALTO, CA 94303

8 9 10 11 12 13 14-MA-95

CONTENTS

(over)

More About Roots and Radicals

Functions

INTRODUCTION

In order to master algebra, most students need a great deal of practice—practice allowing them to discover mathematical patterns, to make generalizations, and to consolidate their mathematical learning—practice that helps them see and understand the workings of algebra. Algebra textbooks, by their very nature, cannot provide the quantity of problems necessary for a beginning algebra course. In order to cover the complete range of problems related to a topic, most textbook exercise sets move very quickly from simple to complex problems, giving only a few of those in-between bread-and-butter problems that students need. As a result, algebra teachers are continually looking for problems to supplement their texts.

About the Series

Developing Skills in Algebra One was created primarily to help teachers in their search for extra algebra problems. The series was not designed to be a classroom text, rather it is a back-up resource containing problems for class examples, chalkboard work, quizzes, test preparation, and extra practice. *Developing Skills in Algebra One* is a four-book series of reproducible worksheets that provides extensive practice in all the work covered in the traditional first high school course in algebra.

Book A starts at the beginning of the school year with exercises in simplifying numerical expressions, and continues through to simple equations in one variable.

Book B includes operations with polynomials, factoring polynomials, solving polynomial equations, and working with rational expressions.

Book C covers ratio, proportion, graphing linear equations, solving systems of linear equations, plus inequalities and absolute value equations.

Book D completes the algebra one course with roots and radicals, quadratic equations, and analysis of quadratic functions.

By design, the books in the *Developing Skills in Algebra One* series are appropriate for use in any algebra one course, whether it is taught in ninth grade, tenth grade, seventh or eighth grade, or in a two-year algebra program. The books also provide review work for second-year algebra students, practice for students studying high school algebra at the college level, and exercises for adults reviewing algebra on their own.

Pick-and-Choose Pages

Book D is the last book in the *Developing Skills in Algebra* series. It contains 100 worksheets with more than 2000 problems that you can duplicate and use with your students. There is no required order for presenting the exercises in this book but, for maximum convenience, the worksheets are arranged sequentially, concept by concept. You may choose to select worksheets from the book as back-up for your algebra lessons. You might assign a worksheet to a single student who needs practice in a specific skill. Or, you may decide to keep certain pages as your own personal resource of problems on a particular topic. The contents in the front of this book and the labels at the top of each worksheet page will help you identify the exercises that best meet your needs.

Paired Worksheets and Exercises

As you glance through this book, you will discover that the worksheets come in pairs; there are at least two parallel worksheets for every concept so that students can learn on one set of problems and practice on the next. Several pairs of worksheets are included for particularly troublesome topics. Each pair of worksheets practices only one or two specific skills (noted at the top left of the pages), carefully sequenced and organized. Most worksheet exercises are also paired, by odds and evens, to allow for two-day assignments or for practice and testing. And, clear simple worksheet instructions along with handwritten samples of the exercises allow students to get right to work with a minimum of fuss.

Checking Work

In order to provide a quantity of problems—
enough problems on a page to make it worth
copying—and in order to give you complete
coverage of algebra topics, we have limited the
amount of workspace allowed for some
exercises. We suggest that you have students
show all their work on separate sheets of
paper, but transfer their answers to the
worksheets. You (or your students) will have a
quick way to check answers as well as access
to the work you must see to diagnose students'
errors of understanding.

You will find answers to every problem in this
book. The answers are located after the
worksheet pages.

About Practice

Practice is an important part of learning, but it's
not the only part. Practice makes sense only
after instruction and demonstrated
understanding. To help students master
algebra, we must aim for a regular and
consistent blend of practice with meaningful
instruction, taking pains to individualize practice
as much as possible. *Developing Skills in
Algebra One* is one tool you can use to achieve
that goal, but it is just a tool. The hard work
and dedication are up to you and your
students.

Developing Skills in

ALGEBRA ONE

Name _____

Date _____ Period _____

Complete each list. Use the diagram at the bottom of the page to help you.

1. List 25 natural numbers. *1, 2, 3, 4, 5, 6, 7, 8, and so on* _____

2. List one whole number that is *not* a natural number. _____

3. List 25 integers that are *not* whole numbers. _____

_____

4. List 25 rational numbers in common fraction form that are *not* integers. _____

_____

5. List 25 rational numbers in decimal form that are terminating decimals but *not* integers. _____

_____

6. List 25 rational numbers in decimal form that are repeating decimals but *not* integers. _____

_____

7. List 25 real numbers that are *not* rational numbers. _____

_____

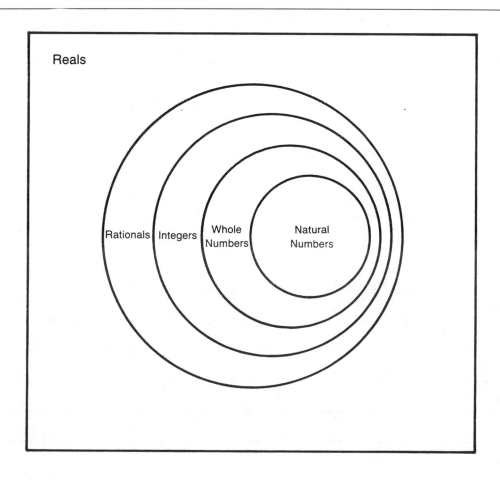

Real Numbers

Use checks to show whether or not the given number belongs in that set.

		natural	whole	integer	rational	irrational
1.	25	✓	✓	✓	✓	
2.	-14					
3.	$\frac{5}{7}$					
4.	0.32					
5.	3.76					
6.	$\sqrt{8}$					
7.	23					
8.	$\sqrt{100}$					
9.	-1.9					
10.	0.47892					
11.	$\sqrt{96}$					
12.	$2\frac{3}{4}$					
13.	$9.3\overline{9}$					
14.	325					
15.	0					
16.	$\sqrt[3]{12}$					
17.	26					
18.	$-\sqrt{17}$					
19.	$13\frac{3}{5}$					
20.	$-2.\overline{378}$					
21.	$\sqrt{64}$					
22.	$5.\overline{5}$					
23.	-12.2					
24.	76					

Developing Skills in Algebra   Book D

Squares and Square Roots

Simplify. Assume variables represent non-negative real numbers.

1. 5^2 *25*

2. 40^2

3. 17^2

4. 9^2

5. 29^2

6. 13^2

7. $\sqrt{1849}$

8. $\sqrt{441}$

9. $\sqrt{1225}$

10. $\sqrt{2401}$

11. $\sqrt{625}$

12. $\sqrt{2500}$

13. $\sqrt{(33)^2}$

14. $\sqrt{(47)^2}$

15. $\sqrt{(-11)^2}$

16. $\sqrt{(-28)^2}$

17. $\sqrt{(36)^2}$

18. $\sqrt{(31)^2}$

19. $\sqrt{(3)^2}\ \sqrt{(22)^2}$

20. $\sqrt{(19)^2}\ \sqrt{(17)^2}$

21. $\sqrt{(-41)^2\,(-8)^2}$

22. $\sqrt{(-75)^2\,(-15)^2}$

23. $\sqrt{\dfrac{4}{25}}$

24. $\sqrt{\dfrac{(36)}{(49)}}$

25. $\sqrt{\dfrac{1}{4}}$

26. $\sqrt{\dfrac{9}{16}}$

27. $\sqrt{1.21}$

28. $\sqrt{17.64}$

29. $\sqrt{13.8384}$

30. $\sqrt{316.84}$

31. $\sqrt{49a^2b^2}$

32. $\sqrt{169c^8d^4}$

33. $\sqrt{256a^{2n}}$

34. $\sqrt{441x^{2r}y^{2r}}$

Squares and Square Roots

Simplify. Assume variables represent non-negative real numbers.

1. 10^2 100

2. 19^2

3. 6^2

4. 33^2

5. 42^2

6. 14^2

7. $\sqrt{2601}$

8. $\sqrt{576}$

9. $\sqrt{1936}$

10. $\sqrt{784}$

11. $\sqrt{1444}$

12. $\sqrt{2304}$

13. $\sqrt{(-21)^2}$

14. $\sqrt{(-15)^2}$

15. $\sqrt{(39)^2}$

16. $\sqrt{(35)^2}$

17. $\sqrt{(52)^2}$

18. $\sqrt{(85)^2}$

19. $\sqrt{(50)^2}\ \sqrt{(16)^2}$

20. $\sqrt{(23)^2}\ \sqrt{(45)^2}$

21. $\sqrt{(-12)^2\,(-32)^2}$

22. $\sqrt{(-60)^2\,(-53)^2}$

23. $\sqrt{\dfrac{1}{64}}$

24. $\sqrt{\dfrac{(16)}{(225)}}$

25. $\sqrt{\dfrac{9}{(25)}}$

26. $\sqrt{\dfrac{(25)}{(49)}}$

27. $\sqrt{12.25}$

28. $\sqrt{31.36}$

29. $\sqrt{39.8161}$

30. $\sqrt{127.69}$

31. $\sqrt{81\,b^4 c^2}$

32. $\sqrt{121\,r^4 s^4 t^2}$

33. $\sqrt{625\,a^{2x}}$

34. $\sqrt{900\,a^{2n} b^{4n}}$

4

Squares and Square Roots

Simplify. Assume variables represent non-negative real numbers.

1. 12^2 144

2. 7^2

3. 18^2

4. 15^2

5. 52^2

6. 36^2

7. $\sqrt{961}$

8. $\sqrt{1681}$

9. $\sqrt{529}$

10. $\sqrt{2209}$

11. $\sqrt{1156}$

12. $\sqrt{3136}$

13. $\sqrt{(29)^2}$

14. $\sqrt{(24)^2}$

15. $\sqrt{(37)^2}$

16. $\sqrt{(25)^2}$

17. $\sqrt{(30)^2}$

18. $\sqrt{(98)^2}$

19. $\sqrt{(43)^2}\ \sqrt{(46)^2}$

20. $\sqrt{(42)^2}\ \sqrt{(20)^2}$

21. $\sqrt{(6)^2\ (27)^2}$

22. $\sqrt{(49)^2\ (38)^2}$

23. $\sqrt{\dfrac{9}{49}}$

24. $\sqrt{\dfrac{(64)}{(289)}}$

25. $\sqrt{\dfrac{16}{(121)}}$

26. $\sqrt{\dfrac{(25)}{(81)}}$

27. $\sqrt{6.76}$

28. $\sqrt{56.25}$

29. $\sqrt{67.7329}$

30. $\sqrt{262.44}$

31. $\sqrt{225\,m^8 n^4}$

32. $\sqrt{484\,x^{14} y^6}$

33. $\sqrt{289\,m^{2n}}$

34. $\sqrt{676\,r^{2x} s^{6x}}$

Squares and Square Roots   Name _____

                                  Date _____ Period _____

Simplify. Assume variables represent non-negative real numbers.

1. 20^2 400 **2.** 30^2

3. 8^2 **4.** 11^2

5. 46^2 **6.** 37^2

7. $\sqrt{1521}$ **8.** $\sqrt{256}$

9. $\sqrt{484}$ **10.** $\sqrt{2025}$

11. $\sqrt{729}$ **12.** $\sqrt{1024}$

13. $\sqrt{(9)^2}$ **14.** $\sqrt{(44)^2}$

15. $\sqrt{(34)^2}$ **16.** $\sqrt{(40)^2}$

17. $\sqrt{(51)^2}$ **18.** $\sqrt{(79)^2}$

19. $\sqrt{(17)^2}\sqrt{(10)^2}$ **20.** $\sqrt{(13)^2}\sqrt{(18)^2}$

21. $\sqrt{(54)^2(72)^2}$ **22.** $\sqrt{(5)^2(48)^2}$

23. $\sqrt{\dfrac{49}{64}}$ **24.** $\sqrt{\dfrac{(9)}{(49)}}$

25. $\sqrt{\dfrac{49}{(100)}}$ **26.** $\sqrt{\dfrac{(64)}{(225)}}$

27. $\sqrt{68.89}$ **28.** $\sqrt{42.25}$

29. $\sqrt{90.6304}$ **30.** $\sqrt{457.96}$

31. $\sqrt{144x^6y^6}$ **32.** $\sqrt{225a^8b^{14}}$

33. $\sqrt{16v^{2a}}$ **34.** $\sqrt{1024c^{6r}d^{10r}}$

Name _____

Date _____ Period _____

Simplify. Assume variables represent non-negative real numbers.

1. $\sqrt{162}$ $9\sqrt{2}$ **2.** $\sqrt{8}$

3. $\sqrt{192}$ **4.** $\sqrt{245}$

5. $\sqrt{3087}$ **6.** $\sqrt{2028}$

7. $\sqrt{98}$ **8.** $\sqrt{720}$

9. $\sqrt{4704}$ **10.** $\sqrt{4400}$

11. $\sqrt{867}$ **12.** $\sqrt{2646}$

13. $\sqrt{700}$ **14.** $\sqrt{1922}$

15. $\sqrt{2352}$ **16.** $\sqrt{320}$

17. $\sqrt{3328}$ **18.** $\sqrt{2904}$

19. $\sqrt{108}$ **20.** $\sqrt{1960}$

21. $\sqrt{8624}$ **22.** $\sqrt{1014}$

23. $\sqrt{3a^2b^4}$ **24.** $\sqrt{2m^4n^2}$

25. $\sqrt{5r^6s^8}$ **26.** $\sqrt{7a^2b^{10}}$

27. $\sqrt{490a^5b^3}$ **28.** $\sqrt{18a^4b^6}$

29. $\sqrt{507a^3}$ **30.** $\sqrt{1210r^3s}$

31. $\sqrt{432c^3d^2}$ **32.** $\sqrt{3125r^7s^4}$

33. $\sqrt{392b^5c^7}$ **34.** $\sqrt{147a^3b^5}$

Radical Expressions

Name _____

Date _____ Period _____

Simplify. Assume variables represent non-negative real numbers.

1. $\sqrt{50}$ $5\sqrt{2}$

2. $\sqrt{363}$

3. $\sqrt{32}$

4. $\sqrt{512}$

5. $\sqrt{726}$

6. $\sqrt{2205}$

7. $\sqrt{5200}$

8. $\sqrt{128}$

9. $\sqrt{6250}$

10. $\sqrt{4205}$

11. $\sqrt{1176}$

12. $\sqrt{1215}$

13. $\sqrt{7497}$

14. $\sqrt{1200}$

15. $\sqrt{1125}$

16. $\sqrt{250}$

17. $\sqrt{10,648}$

18. $\sqrt{3703}$

19. $\sqrt{3072}$

20. $\sqrt{294}$

21. $\sqrt{6750}$

22. $\sqrt{2816}$

23. $\sqrt{2c^4 d^2}$

24. $\sqrt{5r^4 s^4}$

25. $\sqrt{3m^6 n^4}$

26. $\sqrt{10r^6 s^4}$

27. $\sqrt{2925 m^4 n^6}$

28. $\sqrt{500 x^6 y^2}$

29. $\sqrt{1440 c^3}$

30. $\sqrt{735 m^3 n}$

31. $\sqrt{392 a^7 b^5}$

32. $\sqrt{845 c^4 d^3}$

33. $\sqrt{405 r^2 s^9}$

34. $\sqrt{578 m^5 n^5}$

Radical Expressions

Simplify. Assume variables represent non-negative real numbers.

1. $\sqrt{27}$ $3\sqrt{3}$

2. $\sqrt{75}$

3. $\sqrt{567}$

4. $\sqrt{180}$

5. $\sqrt{3456}$

6. $\sqrt{4860}$

7. $\sqrt{1875}$

8. $\sqrt{338}$

9. $\sqrt{1458}$

10. $\sqrt{3610}$

11. $\sqrt{2448}$

12. $\sqrt{722}$

13. $\sqrt{10,935}$

14. $\sqrt{17,100}$

15. $\sqrt{1445}$

16. $\sqrt{208}$

17. $\sqrt{3718}$

18. $\sqrt{3267}$

19. $\sqrt{120}$

20. $\sqrt{1500}$

21. $\sqrt{768}$

22. $\sqrt{1058}$

23. $\sqrt{5m^2n^4}$

24. $\sqrt{3p^4q^2}$

25. $\sqrt{2a^8b^6}$

26. $\sqrt{11x^{10}y^2}$

27. $\sqrt{1352a^6b^2}$

28. $\sqrt{252r^4s^6}$

29. $\sqrt{175x^3}$

30. $\sqrt{1183c^3}$

31. $\sqrt{2645x^7y^7}$

32. $\sqrt{1280p^3q^8}$

33. $\sqrt{288a^{11}b^5}$

34. $\sqrt{1620c^5d^{11}}$

Name _____

Date _____ Period _____

Simplify. Assume variables represent non-negative real numbers.

1. $\sqrt{500}$ $10\sqrt{5}$

2. $\sqrt{486}$

3. $\sqrt{1083}$

4. $\sqrt{1352}$

5. $\sqrt{448}$

6. $\sqrt{3174}$

7. $\sqrt{360}$

8. $\sqrt{3375}$

9. $\sqrt{972}$

10. $\sqrt{3645}$

11. $\sqrt{2268}$

12. $\sqrt{1152}$

13. $\sqrt{2890}$

14. $\sqrt{10,140}$

15. $\sqrt{15,300}$

16. $\sqrt{270}$

17. $\sqrt{5819}$

18. $\sqrt{980}$

19. $\sqrt{4032}$

20. $\sqrt{3750}$

21. $\sqrt{2535}$

22. $\sqrt{4950}$

23. $\sqrt{3r^4s^4}$

24. $\sqrt{5v^4w^2}$

25. $\sqrt{7c^6d^2}$

26. $\sqrt{15a^4b^6}$

27. $\sqrt{243r^4s^2}$

28. $\sqrt{128m^6n^2}$

29. $\sqrt{1859y^3}$

30. $\sqrt{1078a^3b}$

31. $\sqrt{2700m^3n^9}$

32. $\sqrt{1587a^7b^9}$

33. $\sqrt{5070r^7s^5}$

34. $\sqrt{675c^5d^{11}}$

Developing Skills in Algebra   Book D

The Pythagorean Theorem

Name _____

Date _____ Period _____

In a right triangle, if *a* and *b* are the legs and *c* is the hypotenuse, then

$$a^2 + b^2 = c^2$$

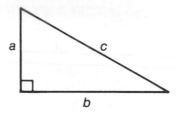

Complete each chart. Assume *a* and *b* are the legs of a right triangle and *c* is the hypotenuse.

	a	*b*	*c*
1.	3	4	5
3.	15	36	
5.	6		10
7.	8		17
9.		20	52
11.		21	29

	a	*b*	*c*
2.	5	12	
4.	25	60	
6.	7		25
8.	15		25
10.		30	34
12.		40	41

Which of the following are Pythagorean Triples? Answer *yes* or *no.*

13. 7, 8, 9   *no*

14. 12, 35, 37

15. 21, 72, 75

16. 11, 45, 60

17. 33, 55, 66

18. 40, 42, 58

19. 30, 40, 50

20. 24, 70, 74

Complete each chart. Assume *a* and *b* are the legs of a right triangle and *c* is the hypotenuse.

	a	*b*	*c*
21.	5	5	$5\sqrt{2}$
23.	8	$4\sqrt{3}$	
25.	$3\sqrt{3}$	9	
27.	7	7	

	a	*b*	*c*
22.	3	$3\sqrt{2}$	
24.	$5\sqrt{6}$	10	
26.	$3\sqrt{5}$	$3\sqrt{2}$	
28.	4	$4\sqrt{7}$	

Developing Skills in Algebra   Book D

The Pythagorean Theorem

Name _____

Date _____ Period _____

In a right triangle, if *a* and *b* are the legs and *c* is the hypotenuse, then

$$a^2 + b^2 = c^2$$

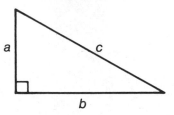

Complete each chart. Assume *a* and *b* are the legs of a right triangle and *c* is the hypotenuse.

	a	*b*	*c*
1.	12	16	20
3.	28	45	
5.	60		87
7.	28		100
9.		11	61
11.		70	74

	a	*b*	*c*
2.	32	60	
4.	10	24	
6.	18		82
8.	11		61
10.		90	102
12.		84	85

Which of the following are Pythagorean Triples? Answer *yes* or *no*.

13. 8, 21, 26   no

14. 18, 24, 30

15. 36, 105, 111

16. 15, 35, 45

17. 24, 45, 51

18. 26, 43, 51

19. 18, 60, 67

20. 80, 84, 116

Complete each chart. Assume *a* and *b* are the legs of a right triangle and *c* is the hypotenuse.

	a	*b*	*c*
21.	4	8	$4\sqrt{5}$
23.	14	$10\sqrt{2}$	
25.	12	8	
27.	15	$3\sqrt{2}$	

	a	*b*	*c*
22.	4	$2\sqrt{3}$	
24.	$4\sqrt{11}$	12	
26.	17	$3\sqrt{6}$	
28.	11	22	

Developing Skills in Algebra   Book D

The Pythagorean Theorem

Name _____

Date _____ Period _____

Solve.

1. The legs of a right triangle measure 33 m and 56 m, respectively. Find the length of the hypotenuse. *Let h = length of hypotenuse*
   $$33^2 + 56^2 = h^2$$
   $$4225 = h^2$$
   $$65 = h$$
   The length of the hypotenuse is 65 cm.

2. The legs of a right triangle measure 36 m and 160 m, respectively. Find the length of the hypotenuse.

3. The legs of a right triangle measure 84 cm and 135 cm, respectively. Find the length of the hypotenuse.

4. The legs of a right triangle measure 100 cm and 105 cm, respectively. Find the length of the hypotenuse.

5. The legs of a right triangle measure 56 km and 105 km, respectively. Find the length of the hypotenuse.

6. The legs of a right triangle measure 51 cm and 68 cm, respectively. Find the length of the hypotenuse.

7. The sides of a triangle measure 16 cm, 63 cm, and 65 cm, respectively. Is the triangle a right triangle?

8. The sides of a triangle measure 39 m, 80 m, and 89 m, respectively. Is the triangle a right triangle?

9. The legs of a right triangle measure $7\sqrt{2}$ m and 21 m, respectively. Find the length of the hypotenuse.

10. The legs of a right triangle measure 18 cm and $9\sqrt{3}$ cm, respectively. Find the length of the hypotenuse.

11. The legs of a right triangle measure 16 km and 8 km, respectively. Find the length of the hypotenuse.

12. The legs of a right triangle measure 15 cm and $3\sqrt{2}$ cm, respectively. Find the length of the hypotenuse.

The Pythagorean Theorem

Name _____

Date _____ Period _____

Solve.

1. The legs of a right triangle measure 40 km and 75 km, respectively. Find the length of the hypotenuse. *Let h = length of hypotenuse*
$$40^2 + 75^2 = h^2$$
$$7225 = h^2$$
$$85 = h$$ *The length of the hypotenuse is 85 km.*

2. The legs of a right triangle measure 27 cm and 120 cm, respectively. Find the length of the hypotenuse.

3. The legs of a right triangle measure 22 m and 120 m, respectively. Find the length of the hypotenuse.

4. The legs of a right triangle measure 48 cm and 55 cm, respectively. Find the length of the hypotenuse.

5. The legs of a right triangle measure 48 m and 140 m, respectively. Find the length of the hypotenuse.

6. The legs of a right triangle measure 35 cm and 120 cm, respectively. Find the length of the hypotenuse.

7. The sides of a triangle measure 28 km, 36 km, and 45 km, respectively. Is the triangle a right triangle?

8. The sides of a triangle measure 20 m, 30 m, and 60 m, respectively. Is the triangle a right triangle?

9. The legs of a right triangle measure 6 cm and $6\sqrt{6}$ cm, respectively. Find the length of the hypotenuse.

10. The legs of a right triangle measure 15 cm and $5\sqrt{11}$ cm, respectively. Find the length of the hypotenuse.

11. The legs of a right triangle measure 25 m and $7\sqrt{7}$ m, respectively. Find the length of the hypotenuse.

12. The legs of a right triangle measure 18 cm and $9\sqrt{3}$ cm, respectively. Find the length of the hypotenuse.

Developing Skills in Algebra   Book D

Radical Expressions

Name _____

Date _____ Period _____

Simplify. Assume all radicands are non-negative numbers.

1. $\sqrt{(x+y)^2}$ $x + y$

2. $\sqrt{(p-q)^2}$

3. $\sqrt{(2a+b)}\,\sqrt{(2a+b)}$

4. $\sqrt{(3c+2d)}\,\sqrt{(3c+2d)}$

5. $\sqrt{m^2+2mn+n^2}$

6. $\sqrt{r^2-2rs+s^2}$

7. $\sqrt{9x^2-12x+4}$

8. $\sqrt{16x^2+40x+25}$

9. $\sqrt{25x^2+20x+4}$

10. $\sqrt{36x^2-60x+25}$

11. $\sqrt{(x+3)^4}$

12. $\sqrt{(x-5)^6}$

13. $\sqrt{(5-7x)^8}$

14. $\sqrt{(8-3x)^4}$

15. $\sqrt{(2x+3)^6}$

16. $\sqrt{(4x-7)^3}$

17. $\sqrt{(2x-3y)^5}\,\sqrt{(2x-3y)^3}$

18. $\sqrt{(4x-9y)^3}\,\sqrt{(4x-9y)^3}$

19. $\sqrt{(x^2+3x+2)^2}$

20. $\sqrt{(x^2-2x-15)^2}$

21. $\sqrt{(8x^2-10x+3)^2}$

22. $\sqrt{(6x^2-11x+10)^2}$

23. $\sqrt{16(2x-7)^2}$

24. $\sqrt{49(5x+9)^2}$

25. $\sqrt{81(3x+1)^4}$

26. $\sqrt{100(2x-13)^6}$

27. $\sqrt{48(8x-1)^2}$

28. $\sqrt{72(10x-3)^2}$

29. $\sqrt{162x^3(3x+1)^3}$

30. $\sqrt{245x^5(2x+13)^5}$

31. $\sqrt{252x^2y^3(5x-7y)^2}$

32. $\sqrt{648x^5y(7x+3y)^3}$

33. $\sqrt{605x^4y(10x+3y)^3}$

34. $\sqrt{338xy^6(5x-17y)^2}$

Radical Expressions

Name _____

Date _____ Period _____

Simplify. Assume all radicands are non-negative numbers.

1. $\sqrt{(m+n)^2}$ *m + n*

2. $\sqrt{(r-s)^2}$

3. $\sqrt{2a-b}\ \sqrt{2a-b}$

4. $\sqrt{3x+y}\ \sqrt{3x+y}$

5. $\sqrt{x^2+10x+25}$

6. $\sqrt{m^2-8m+16}$

7. $\sqrt{9x^2-30x+25}$

8. $\sqrt{16x^2+24x+9}$

9. $\sqrt{49x^2-28x+4}$

10. $\sqrt{64x^2-48x+9}$

11. $\sqrt{(x-7)^6}$

12. $\sqrt{(x+9)^4}$

13. $\sqrt{(10-3x)^4}$

14. $\sqrt{(7-13x)^8}$

15. $\sqrt{(4x+17)^3}$

16. $\sqrt{(3x-19)^5}$

17. $\sqrt{(2x+1)^5}\ \sqrt{(2x+1)}$

18. $\sqrt{(3x-7)^3}\ \sqrt{(3x-7)^3}$

19. $\sqrt{(2x^2+11x+5)^2}$

20. $\sqrt{(6x^2-5x-4)^2}$

21. $\sqrt{(8x^2+25x-3)^2}$

22. $\sqrt{(12x^2-28x+15)^2}$

23. $\sqrt{25(3x+5)^2}$

24. $\sqrt{36(7x-3)^2}$

25. $\sqrt{169(5-2x)^4}$

26. $\sqrt{225(10x^2-9x)^2}$

27. $\sqrt{175(3x+8)^2}$

28. $\sqrt{405(11+3x)^2}$

29. $\sqrt{54x^9(5x+3)^5}$

30. $\sqrt{567x^5(2x-11)^4}$

31. $\sqrt{275x^4y^5(3x-7y)^3}$

32. $\sqrt{1445x^6y(2x+9y)^7}$

33. $\sqrt{722x^6y(8x+3y)^5}$

34. $\sqrt{192x^7y^4(3x-10y)^3}$

16

Name _____

Date _____ Period _____

Multiply. Write answers in simplified radical form. Assume variables represent non-negative real numbers.

1. $\sqrt{2}\,\sqrt{3}$ $\quad\sqrt{6}$

2. $\sqrt{2}\,\sqrt{2}$

3. $\sqrt{5}\,\sqrt{6}$

4. $\sqrt{6}\,\sqrt{8}$

5. $\sqrt{8}\,\sqrt{8}$

6. $\sqrt{2}\,\sqrt{6}$

7. $3\sqrt{2}\,\sqrt{10}$

8. $\sqrt{5}\,\sqrt{18}$

9. $\sqrt{3}\,\sqrt{18}$

10. $5\sqrt{5}\,\sqrt{20}$

11. $\sqrt{15}\,\sqrt{30}$

12. $\sqrt{2}\,\sqrt{24}$

13. $\sqrt{18}\,\sqrt{45}$

14. $2\sqrt{23}\,\,7\sqrt{40}$

15. $2\sqrt{27}\,\,3\sqrt{30}$

16. $\sqrt{3}\,\sqrt{21}$

17. $\sqrt{6}\,\sqrt{8}\,\sqrt{10}$

18. $\sqrt{10}\,\sqrt{14}\,\sqrt{21}$

19. $\sqrt{18}\,\sqrt{23}\,\sqrt{33}$

20. $\sqrt{20}\,\sqrt{26}\,\sqrt{39}$

21. $\sqrt{18a^2b^3}\,\sqrt{27a^2b^3}$

22. $\sqrt{32x^2y^2}\,\sqrt{18xy^3}$

23. $4\sqrt{22c^3d^2}\,\,3\sqrt{28c^3d^2}$

24. $\sqrt{34pq^3}\,\sqrt{51p^2q^3}$

25. $\sqrt{27r^5s^2}\,\sqrt{40r^3s^2}$

26. $2\sqrt{38a^3b^2}\,\,3\sqrt{40ab^3}$

27. $\sqrt{33x^3y^3}\,\sqrt{99xy^2}$

28. $\sqrt{40m^2n^2}\,\sqrt{60m^3n^2}$

29. $\sqrt{11a^2}\,\sqrt{22a^2}\,\sqrt{16a^3}$

30. $\sqrt{15x^3}\,\sqrt{18x^2}\,\sqrt{12x^2}$

31. $5\sqrt{17y^2}\,\,2\sqrt{24y^3}\,\sqrt{34y^2}$

32. $\sqrt{24m^2}\,\,4\sqrt{27m^3}\,\,3\sqrt{32m^3}$

33. $\sqrt{15r^3}\,\sqrt{20r^5}\,\sqrt{35r^3}$

34. $\sqrt{25t^5}\,\sqrt{15t^3}\,\sqrt{40t^2}$

Multiplication of Radical Expressions

Name _____

Date _____ Period _____

Multiply. Write answers in simplified radical form. Assume variables represent non-negative real numbers.

1. $\sqrt{2}\,\sqrt{4}$ $2\sqrt{2}$

2. $\sqrt{2}\,\sqrt{7}$

3. $\sqrt{7}\,\sqrt{8}$

4. $\sqrt{2}\,\sqrt{5}$

5. $2\sqrt{3}\,\sqrt{5}$

6. $\sqrt{5}\;3\sqrt{8}$

7. $\sqrt{9}\,\sqrt{9}$

8. $\sqrt{2}\,\sqrt{12}$

9. $5\sqrt{10}\;2\sqrt{12}$

10. $\sqrt{5}\,\sqrt{15}$

11. $\sqrt{8}\,\sqrt{10}$

12. $6\sqrt{15}\;5\sqrt{35}$

13. $\sqrt{24}\,\sqrt{38}$

14. $\sqrt{26}\,\sqrt{39}$

15. $\sqrt{22}\,\sqrt{24}$

16. $\sqrt{32}\,\sqrt{10}$

17. $3\sqrt{12}\;4\sqrt{15}\,\sqrt{18}$

18. $2\sqrt{18}\,\sqrt{20}\;5\sqrt{24}$

19. $\sqrt{8}\,\sqrt{10}\,\sqrt{12}$

20. $\sqrt{26}\,\sqrt{28}\,\sqrt{30}$

21. $2\sqrt{22x^2y}\;3\sqrt{33xy^2}$

22. $\sqrt{40a^2b^3}\,\sqrt{42a^2b^3}$

23. $\sqrt{20m^3n^2}\,\sqrt{24m^2n^3}$

24. $\sqrt{42r^3s}\,\sqrt{48r^2s^3}$

25. $\sqrt{40x^3y^3}\,\sqrt{48x^2y^2}$

26. $4\sqrt{44a^3b^2}\;5\sqrt{56a^5b^2}$

27. $\sqrt{44r^3s^2}\,\sqrt{66r^3s}$

28. $\sqrt{33m^5n^3}\,\sqrt{63m^3n^2}$

29. $3\sqrt{13a^4}\;4\sqrt{26a^3}\,\sqrt{27a^2}$

30. $\sqrt{12y^3}\,\sqrt{18y^2}\,\sqrt{27y^2}$

31. $\sqrt{19y^3}\,\sqrt{57y^2}\,\sqrt{6y^3}$

32. $\sqrt{45m^3}\,\sqrt{50m^2}\,\sqrt{15m^3}$

33. $\sqrt{18a^2}\,\sqrt{30a^3}\,\sqrt{45a^3}$

34. $3\sqrt{18t^4}\;7\sqrt{27t^3}\,\sqrt{51t^2}$

    Developing Skills in Algebra   Book D

Multiplication of Radical Expressions

Multiply. Write answers in simplified radical form. Assume variables represent non-negative real numbers.

1. $\sqrt{2}\,\sqrt{8}$ 4

2. $\sqrt{7}\,\sqrt{7}$

3. $\sqrt{3}\,\sqrt{6}$

4. $\sqrt{9}\,\sqrt{8}$

5. $\sqrt{3}\,\sqrt{12}$

6. $\sqrt{3}\,\sqrt{7}$

7. $3\sqrt{5}\,\sqrt{10}$

8. $\sqrt{2}\,\,7\sqrt{15}$

9. $\sqrt{18}\,\sqrt{24}$

10. $\sqrt{2}\,\sqrt{18}$

11. $3\sqrt{2}\,\sqrt{26}$

12. $\sqrt{5}\,\,5\sqrt{24}$

13. $\sqrt{26}\,\sqrt{48}$

14. $\sqrt{32}\,\sqrt{40}$

15. $\sqrt{3}\,\sqrt{27}$

16. $\sqrt{18}\,\sqrt{60}$

17. $\sqrt{20}\,\sqrt{24}\,\sqrt{30}$

18. $\sqrt{18}\,\sqrt{30}\,\sqrt{45}$

19. $6\sqrt{10}\,\,2\sqrt{18}\,\sqrt{22}$

20. $3\sqrt{28}\,\sqrt{32}\,\,4\sqrt{40}$

21. $\sqrt{24r^3s^3}\,\sqrt{50r^2s^3}$

22. $\sqrt{33p^3q^3}\,\sqrt{42p^2q}$

23. $2\sqrt{27a^3b^2}\,\,3\sqrt{39a^3b}$

24. $\sqrt{34r^3s}\,\sqrt{50rs^3}$

25. $\sqrt{20x^5y^3}\,\sqrt{32x^2y^3}$

26. $\sqrt{40m^2n^3}\,\sqrt{50m^3n^3}$

27. $\sqrt{27a^3b^3}\,\sqrt{33a^2b^3}$

28. $7\sqrt{42x^4y^3}\,\,2\sqrt{56x^5y^3}$

29. $4\sqrt{15r^3}\,\sqrt{30r^2}\,\,3\sqrt{38r^3}$

30. $\sqrt{21y^2}\,\sqrt{42y^2}\,\sqrt{14y^2}$

31. $\sqrt{21a^4}\,\sqrt{24a^2}\,\sqrt{27a^2}$

32. $\sqrt{18p^3}\,\sqrt{30p^2}\,\sqrt{40p^3}$

33. $\sqrt{26a^3}\,\sqrt{52a^3}\,\sqrt{32a^3}$

34. $\sqrt{15z^2}\,\sqrt{21z^3}\,\sqrt{35z^2}$

19

Multiplication of Radical Expressions

Name _____

Date _____ Period _____

Multiply. Write answers in simplified radical form. Assume variables represent non-negative real numbers.

1. $\sqrt{3}\ \sqrt{3}$ *3*

2. $\sqrt{6}\ \sqrt{7}$

3. $\sqrt{8}\ \sqrt{12}$

4. $\sqrt{3}\ \sqrt{8}$

5. $\sqrt{5}\ \sqrt{9}$

6. $\sqrt{3}\ \sqrt{15}$

7. $5\sqrt{8}\ 9\sqrt{18}$

8. $\sqrt{8}\ \sqrt{32}$

9. $\sqrt{2}\ \sqrt{21}$

10. $5\sqrt{18}\ 7\sqrt{30}$

11. $\sqrt{22}\ \sqrt{24}$

12. $\sqrt{27}\ \sqrt{30}$

13. $\sqrt{8}\ \sqrt{26}$

14. $\sqrt{32}\ \sqrt{60}$

15. $2\sqrt{3}\ 3\sqrt{33}$

16. $\sqrt{34}\ \sqrt{40}$

17. $\sqrt{12}\ \sqrt{20}\ \sqrt{24}$

18. $6\sqrt{10}\ 3\sqrt{20}\ \sqrt{27}$

19. $\sqrt{30}\ \sqrt{32}\ \sqrt{38}$

20. $\sqrt{40}\ \sqrt{54}\ \sqrt{50}$

21. $\sqrt{27p^3q^2}\ \sqrt{50p^2q^3}$

22. $\sqrt{40s^3t^3}\ \sqrt{44s^2t}$

23. $5\sqrt{32x^3y^2}\ 2\sqrt{44x^3y^2}$

24. $5\sqrt{42r^2s}\ 2\sqrt{35r^4s^3}$

25. $\sqrt{33a^3b^5}\ \sqrt{60a^3b^3}$

26. $\sqrt{42m^3n^3}\ \sqrt{63m^5n}$

27. $7\sqrt{20x^3y^3}\ 3\sqrt{40x^2y}$

28. $\sqrt{38c^2d^3}\ \sqrt{57c^2d^3}$

29. $\sqrt{7a^3}\ \sqrt{21a^3}\ \sqrt{33a^3}$

30. $6\sqrt{23y^2}\ \sqrt{46y^3}\ 3\sqrt{8y^3}$

31. $\sqrt{18c^2}\ \sqrt{24c^3}\ \sqrt{14c^2}$

32. $\sqrt{26x^3}\ \sqrt{39x^5}\ \sqrt{12x^3}$

33. $\sqrt{21r^3}\ \sqrt{35r^3}\ \sqrt{33r^2}$

34. $\sqrt{42z^2}\ \sqrt{21z^2}\ \sqrt{33z^3}$

Multiplication of Radical Expressions

Name _____

Date _____ Period _____

Multiply. Simplify all radicals. Assume all radicands are non-negative numbers.

1. $\sqrt{2}\left(3x + 2y\right)$ $3x\sqrt{2} + 2y\sqrt{2}$

2. $\sqrt{3}\left(2x - 5y\right)$

3. $\sqrt{7}\left(4a - 9b\right)$

4. $\sqrt{11}\left(11m + 3n\right)$

5. $\sqrt{8}\left(2x^2 + 7x - 1\right)$

6. $\sqrt{27}\left(3x^2 - 5x + 2\right)$

7. $\sqrt{75}\left(x^2 - 5x + 3\right)$

8. $\sqrt{12}\left(x^2 - 9x - 1\right)$

9. $\sqrt{45x^2}\left(x^2 - 10x - 2\right)$

10. $\sqrt{48x^4}\left(x^2 - 7x + 3\right)$

11. $\sqrt{162x^8}\left(x^3 + 7x - 5\right)$

12. $\sqrt{63x^6}\left(x^3 - 2x + 5\right)$

13. $\sqrt{3}\left(\sqrt{3} + \sqrt{7}\right)$

14. $\sqrt{5}\left(\sqrt{7} + \sqrt{5}\right)$

15. $\sqrt{11}\left(\sqrt{5} + \sqrt{11}\right)$

16. $\sqrt{13}\left(\sqrt{13} + \sqrt{3}\right)$

17. $\sqrt{15}\left(\sqrt{3} + \sqrt{5}\right)$

18. $\sqrt{20}\left(\sqrt{6} + \sqrt{5}\right)$

19. $\sqrt{6}\left(\sqrt{10} + \sqrt{15}\right)$

20. $\sqrt{14}\left(\sqrt{7} + \sqrt{10}\right)$

21. $a\sqrt{5}\left(a\sqrt{2} - b\sqrt{3}\right)$

22. $x\sqrt{7}\left(x\sqrt{3} + y\sqrt{5}\right)$

23. $m\sqrt{3}\left(m\sqrt{5} + n\sqrt{7}\right)$

24. $z\sqrt{2}\left(y\sqrt{3} - z\sqrt{7}\right)$

25. $x\sqrt{2}\left(x\sqrt{14} - y\sqrt{18}\right)$

26. $m\sqrt{3}\left(m\sqrt{6} + n\sqrt{12}\right)$

27. $r\sqrt{7}\left(r\sqrt{14} + s\sqrt{21}\right)$

28. $t\sqrt{11}\left(s\sqrt{22} - t\sqrt{33}\right)$

29. $\sqrt{27x^3}\left(x\sqrt{3} - y\sqrt{2}\right)$

30. $\sqrt{48y}\left(x\sqrt{2} - y\sqrt{3}\right)$

31. $\sqrt{68m^5}\left(3m\sqrt{17m} - 2n\sqrt{2n}\right)$

32. $\sqrt{54a}\left(a\sqrt{3a} + 4b\sqrt{2b}\right)$

33. $\sqrt{80s^3}\left(5s\sqrt{10s} + 2t\sqrt{2t}\right)$

34. $\sqrt{273x}\left(x\sqrt{3x} + 2y\sqrt{5y}\right)$

Developing Skills in Algebra   Book D

Multiplication of Radical Expressions

Name _____

Date _____ Period _____

Multiply. Simplify all radicals. Assume all radicands are non-negative numbers.

1. $\sqrt{5}\left(7a + 3b\right)$ $7a\sqrt{5} + 3b\sqrt{5}$

2. $\sqrt{2}\left(5x - 4y\right)$

3. $\sqrt{13}\left(5m - 2n\right)$

4. $\sqrt{17}\left(6r + 2t\right)$

5. $\sqrt{80}\left(3x^2 - 5x + 7\right)$

6. $\sqrt{32}\left(2x^2 + 7x - 5\right)$

7. $\sqrt{125}\left(5x^2 - 3x + 1\right)$

8. $\sqrt{243}\left(x^2 - 7x - 3\right)$

9. $\sqrt{72x^2}\left(x^2 - 3x + 7\right)$

10. $\sqrt{242x^4}\left(x^2 + 5x + 3\right)$

11. $\sqrt{112x^6}\left(2x^2 + 5x - 3\right)$

12. $\sqrt{175x^8}\left(5x^2 - 4x + 1\right)$

13. $\sqrt{2}\left(\sqrt{5} + \sqrt{2}\right)$

14. $\sqrt{7}\left(\sqrt{3} - \sqrt{7}\right)$

15. $\sqrt{5}\left(\sqrt{11} - \sqrt{5}\right)$

16. $\sqrt{13}\left(\sqrt{2} + \sqrt{13}\right)$

17. $\sqrt{33}\left(\sqrt{5} + \sqrt{11}\right)$

18. $\sqrt{39}\left(\sqrt{3} + \sqrt{13}\right)$

19. $\sqrt{34}\left(\sqrt{2} + \sqrt{17}\right)$

20. $\sqrt{10}\left(\sqrt{5} - \sqrt{2}\right)$

21. $x\sqrt{7}\left(x\sqrt{3} - y\sqrt{5}\right)$

22. $c\sqrt{5}\left(b\sqrt{2} + c\sqrt{3}\right)$

23. $a\sqrt{2}\left(a\sqrt{7} + b\sqrt{3}\right)$

24. $y\sqrt{11}\left(x\sqrt{2} - y\sqrt{5}\right)$

25. $x\sqrt{3}\left(x\sqrt{15} - y\sqrt{30}\right)$

26. $x\sqrt{7}\left(x\sqrt{3} - y\sqrt{14}\right)$

27. $t\sqrt{5}\left(s\sqrt{10} + t\sqrt{15}\right)$

28. $v\sqrt{13}\left(v\sqrt{26} - w\sqrt{39}\right)$

29. $\sqrt{176x}\left(x\sqrt{2} - y\sqrt{3}\right)$

30. $\sqrt{45y^3}\left(y\sqrt{3} - z\sqrt{5}\right)$

31. $\sqrt{128a^3}\left(2a\sqrt{12a} - 5b\sqrt{10b}\right)$

32. $\sqrt{162c^2}\left(3c\sqrt{6c} - 7d\sqrt{2d}\right)$

33. $\sqrt{150r^5}\left(r\sqrt{10r} + 2s\sqrt{12s}\right)$

34. $\sqrt{48p}\left(5p\sqrt{3p} - 8q\sqrt{2q}\right)$

22

Division of Rational Expressions

Simplify.

1. $\sqrt{\dfrac{48}{3}}$ 4

2. $\sqrt{\dfrac{98}{2}}$

3. $\sqrt{\dfrac{405}{5}}$

4. $\sqrt{\dfrac{128}{2}}$

5. $\sqrt{\dfrac{1}{2}}$

6. $\sqrt{\dfrac{2}{3}}$

7. $\sqrt{\dfrac{4}{5}}$

8. $\sqrt{\dfrac{9}{10}}$

9. $\sqrt{\dfrac{16}{3}}$

10. $\sqrt{\dfrac{64}{7}}$

11. $\sqrt{\dfrac{49}{2}}$

12. $\sqrt{\dfrac{81}{5}}$

13. $\dfrac{\sqrt{15}}{\sqrt{5}}$

14. $\dfrac{\sqrt{26}}{\sqrt{2}}$

15. $\dfrac{\sqrt{30}}{\sqrt{6}}$

16. $\dfrac{\sqrt{42}}{\sqrt{7}}$

17. $\dfrac{\sqrt{15}}{\sqrt{7}}$

18. $\dfrac{\sqrt{21}}{\sqrt{5}}$

19. $\dfrac{\sqrt{11}}{\sqrt{2}}$

20. $\dfrac{\sqrt{17}}{\sqrt{3}}$

21. $\dfrac{\sqrt{28}}{\sqrt{6}}$

22. $\dfrac{\sqrt{18}}{\sqrt{10}}$

23. $\dfrac{\sqrt{48}}{\sqrt{14}}$

24. $\dfrac{\sqrt{75}}{\sqrt{21}}$

25. $\dfrac{\sqrt{98}}{\sqrt{22}}$

26. $\dfrac{\sqrt{162}}{\sqrt{34}}$

Developing Skills in Algebra   Book D

Division of Rational Expressions

Simplify.

1. $\sqrt{\dfrac{245}{5}}$ 7

2. $\sqrt{\dfrac{288}{2}}$

3. $\sqrt{\dfrac{112}{7}}$

4. $\sqrt{\dfrac{363}{3}}$

5. $\sqrt{\dfrac{3}{5}}$

6. $\sqrt{\dfrac{5}{7}}$

7. $\sqrt{\dfrac{7}{8}}$

8. $\sqrt{\dfrac{3}{10}}$

9. $\sqrt{\dfrac{100}{7}}$

10. $\sqrt{\dfrac{144}{5}}$

11. $\sqrt{\dfrac{121}{13}}$

12. $\sqrt{\dfrac{169}{11}}$

13. $\dfrac{\sqrt{28}}{\sqrt{7}}$

14. $\dfrac{\sqrt{51}}{\sqrt{17}}$

15. $\dfrac{\sqrt{95}}{\sqrt{5}}$

16. $\dfrac{\sqrt{110}}{\sqrt{11}}$

17. $\dfrac{\sqrt{33}}{\sqrt{10}}$

18. $\dfrac{\sqrt{42}}{\sqrt{55}}$

19. $\dfrac{\sqrt{29}}{\sqrt{3}}$

20. $\dfrac{\sqrt{38}}{\sqrt{17}}$

21. $\dfrac{\sqrt{112}}{\sqrt{22}}$

22. $\dfrac{\sqrt{275}}{\sqrt{35}}$

23. $\dfrac{\sqrt{180}}{\sqrt{57}}$

24. $\dfrac{\sqrt{567}}{\sqrt{14}}$

25. $\dfrac{\sqrt{147}}{\sqrt{6}}$

26. $\dfrac{\sqrt{605}}{\sqrt{10}}$

Developing Skills in Algebra   Book D

Division of Rational Expressions

Name _____

Date _____ Period _____

Simplify. Assume variables represent non-negative numbers.

1. $\sqrt{\dfrac{1}{2}}$ $\dfrac{\sqrt{2}}{2}$

2. $\sqrt{\dfrac{1}{13}}$

3. $\sqrt{\dfrac{961}{15}}$

4. $\sqrt{\dfrac{1444}{7}}$

5. $\sqrt{\dfrac{27}{5}}$

6. $\dfrac{1}{\sqrt{3}}$

7. $\sqrt{\dfrac{15}{8}}$

8. $\sqrt{\dfrac{75}{13}}$

9. $\dfrac{1}{\sqrt{5}}$

10. $\sqrt{\dfrac{35}{24}}$

11. $\sqrt{\dfrac{58}{6}}$

12. $\sqrt{\dfrac{135}{21}}$

13. $\dfrac{\sqrt{19}}{\sqrt{54}}$

14. $\dfrac{\sqrt{13}}{\sqrt{27}}$

15. $\dfrac{\sqrt{867}}{\sqrt{125}}$

16. $\dfrac{1}{\sqrt{8}}$

17. $\sqrt{\dfrac{5\,c^2\,d^2}{e}}$

18. $\sqrt{\dfrac{7\,ab^3}{c}}$

19. $\sqrt{\dfrac{1}{x}}$

20. $\sqrt{\dfrac{3}{y}}$

21. $\sqrt{\dfrac{11}{y^3}}$

22. $\sqrt{\dfrac{1}{x^3}}$

23. $\dfrac{\sqrt{28\,p}}{\sqrt{7q}}$

24. $\dfrac{\sqrt{45\,d}}{\sqrt{5e}}$

25. $\dfrac{\sqrt{98\,m}}{\sqrt{3n^3}}$

26. $\dfrac{\sqrt{72\,x^2}}{\sqrt{5y^3}}$

Developing Skills in Algebra   Book D

Division of Rational Expressions

Name _____

Date _____ Period _____

Simplify. Assume variables represent non-negative numbers.

1. $\sqrt{\dfrac{1}{13}}$ $\dfrac{\sqrt{13}}{13}$

2. $\sqrt{\dfrac{1}{7}}$

3. $\sqrt{\dfrac{1296}{11}}$

4. $\sqrt{\dfrac{1521}{5}}$

5. $\sqrt{\dfrac{128}{3}}$

6. $\dfrac{1}{\sqrt{343}}$

7. $\sqrt{\dfrac{125}{7}}$

8. $\sqrt{\dfrac{21}{8}}$

9. $\dfrac{1}{\sqrt{6}}$

10. $\sqrt{\dfrac{48}{15}}$

11. $\sqrt{\dfrac{92}{26}}$

12. $\sqrt{\dfrac{243}{5}}$

13. $\dfrac{\sqrt{13}}{\sqrt{112}}$

14. $\dfrac{\sqrt{15}}{\sqrt{175}}$

15. $\dfrac{\sqrt{338}}{\sqrt{8}}$

16. $\dfrac{1}{\sqrt{729}}$

17. $\sqrt{\dfrac{14\,ar}{t}}$

18. $\sqrt{\dfrac{12\,a^2\,c}{b}}$

19. $\sqrt{\dfrac{1}{z}}$

20. $\sqrt{\dfrac{5}{q}}$

21. $\sqrt{\dfrac{7}{m^3}}$

22. $\sqrt{\dfrac{10}{y^3}}$

23. $\dfrac{\sqrt{44\,z}}{\sqrt{3y}}$

24. $\dfrac{\sqrt{75\,r}}{\sqrt{3t}}$

25. $\dfrac{\sqrt{99\,a}}{\sqrt{5b^4}}$

26. $\dfrac{\sqrt{56\,t^2}}{\sqrt{3r^5}}$

     Developing Skills in Algebra   Book D

Division of Rational Expressions

Name _____

Date _____ Period _____

Divide. Write your answers in simplified radical form.

1. $\dfrac{6\sqrt{3} - 7\sqrt{6}}{\sqrt{3}}$ $6 - 7\sqrt{2}$

2. $\dfrac{5\sqrt{2} + 8\sqrt{10}}{\sqrt{2}}$

3. $\dfrac{\sqrt{5} + 8\sqrt{15}}{\sqrt{5}}$

4. $\dfrac{3\sqrt{7} + \sqrt{14}}{\sqrt{7}}$

5. $\dfrac{\sqrt{8} + 3\sqrt{12}}{\sqrt{2}}$

6. $\dfrac{4\sqrt{27} - 5\sqrt{45}}{\sqrt{3}}$

7. $\dfrac{\sqrt{98} - 6\sqrt{128}}{\sqrt{2}}$

8. $\dfrac{9\sqrt{48} - 7\sqrt{147}}{\sqrt{3}}$

9. $\dfrac{8 - \sqrt{3}}{\sqrt{3}}$

10. $\dfrac{4\sqrt{5} + 10}{\sqrt{5}}$

11. $\dfrac{\sqrt{7} + 9}{7}$

12. $\dfrac{12 - \sqrt{2}}{\sqrt{2}}$

13. $\dfrac{\sqrt{3} + \sqrt{6}}{\sqrt{2}}$

14. $\dfrac{\sqrt{5} - \sqrt{2}}{\sqrt{5}}$

15. $\dfrac{3\sqrt{7} - 2\sqrt{5}}{\sqrt{5}}$

16. $\dfrac{8\sqrt{11} + 3\sqrt{3}}{\sqrt{3}}$

17. $\dfrac{2\sqrt{8} + 3\sqrt{27}}{\sqrt{5}}$

18. $\dfrac{4\sqrt{343} - 6\sqrt{125}}{\sqrt{3}}$

19. $\dfrac{5\sqrt{48} - 2\sqrt{54}}{\sqrt{7}}$

20. $\dfrac{6\sqrt{63} + 7\sqrt{80}}{\sqrt{2}}$

21. $\dfrac{3\sqrt{32} + 7\sqrt{108}}{\sqrt{3}}$

22. $\dfrac{5\sqrt{175} + 2\sqrt{432}}{\sqrt{5}}$

23. $\dfrac{2\sqrt{52} - 3\sqrt{90}}{\sqrt{13}}$

24. $\dfrac{7\sqrt{162} - 5\sqrt{150}}{\sqrt{2}}$

25. $\dfrac{21\sqrt{288} + 13\sqrt{845}}{\sqrt{7}}$

26. $\dfrac{4\sqrt{275} + 3\sqrt{147}}{\sqrt{3}}$

27

Division of Rational Expressions

Name _____

Date _____ Period _____

Divide. Write your answers in simplified radical form.

1. $\dfrac{8\sqrt{2} - 3\sqrt{14}}{\sqrt{2}}$ $8 - 3\sqrt{7}$

2. $\dfrac{9\sqrt{5} + 3\sqrt{15}}{\sqrt{5}}$

3. $\dfrac{3\sqrt{14} + 2\sqrt{21}}{\sqrt{7}}$

4. $\dfrac{9\sqrt{22} - 4\sqrt{33}}{\sqrt{11}}$

5. $\dfrac{3\sqrt{98} + 6\sqrt{12}}{\sqrt{2}}$

6. $\dfrac{4\sqrt{48} - 3\sqrt{15}}{\sqrt{3}}$

7. $\dfrac{\sqrt{52} - 2\sqrt{117}}{\sqrt{13}}$

8. $\dfrac{2\sqrt{27} + \sqrt{33}}{\sqrt{3}}$

9. $\dfrac{4 + \sqrt{10}}{\sqrt{10}}$

10. $\dfrac{15\sqrt{7} + 10}{\sqrt{7}}$

11. $\dfrac{13\sqrt{11} + 3}{\sqrt{11}}$

12. $\dfrac{9 - 3\sqrt{6}}{\sqrt{6}}$

13. $\dfrac{\sqrt{5} + \sqrt{10}}{\sqrt{2}}$

14. $\dfrac{\sqrt{14} + \sqrt{15}}{\sqrt{7}}$

15. $\dfrac{4\sqrt{15} + 3\sqrt{2}}{\sqrt{3}}$

16. $\dfrac{3\sqrt{26} + \sqrt{5}}{\sqrt{13}}$

17. $\dfrac{7\sqrt{27} + 8\sqrt{20}}{\sqrt{2}}$

18. $\dfrac{5\sqrt{32} - 3\sqrt{28}}{\sqrt{3}}$

19. $\dfrac{3\sqrt{125} - 7\sqrt{98}}{\sqrt{7}}$

20. $\dfrac{4\sqrt{343} + 2\sqrt{405}}{\sqrt{11}}$

21. $\dfrac{5\sqrt{80} + 3\sqrt{162}}{\sqrt{8}}$

22. $\dfrac{12\sqrt{60} + 3\sqrt{200}}{\sqrt{7}}$

23. $\dfrac{6\sqrt{240} - 8\sqrt{108}}{\sqrt{5}}$

24. $\dfrac{3\sqrt{54} - 7\sqrt{112}}{\sqrt{27}}$

25. $\dfrac{2\sqrt{176} + 3\sqrt{117}}{\sqrt{2}}$

26. $\dfrac{4\sqrt{28} - 3\sqrt{135}}{\sqrt{3}}$

Developing Skills in Algebra   Book D

Factoring Radical Expressions

Name _____

Date _____ Period _____

Factor each expression. Radical expressions should be in simplified form. Assume all variables represent non-negative numbers.

1. $7x\sqrt{5} + 2y\sqrt{5}$ $\sqrt{5}(7x + 2y)$

2. $3c\sqrt{2} + 5d\sqrt{2}$

3. $11y\sqrt{3} - 7z\sqrt{3}$

4. $4t\sqrt{11} - 7s\sqrt{11}$

5. $2a\sqrt{3b} + 4c\sqrt{3b}$

6. $6x\sqrt{5y} - 9z\sqrt{5y}$

7. $12t\sqrt{5r} - 8s\sqrt{5r}$

8. $6z\sqrt{3z} + 14y\sqrt{3z}$

9. $x^2\sqrt{2} - 7x\sqrt{2} + \sqrt{2}$

10. $x^2\sqrt{3} - 5x\sqrt{3} + 2\sqrt{3}$

11. $x^2\sqrt{5} + 9x\sqrt{5} + 3\sqrt{5}$

12. $x^2\sqrt{7} + 2x\sqrt{7} - 9\sqrt{7}$

13. $x\sqrt{27} + y\sqrt{12}$

14. $a\sqrt{45} - b\sqrt{80}$

15. $r\sqrt{128} + t\sqrt{50}$

16. $c\sqrt{147} - d\sqrt{243}$

17. $p\sqrt{45q^2} - q\sqrt{125q^2}$

18. $x\sqrt{112b^2} - x\sqrt{343a^2}$

19. $c\sqrt{28t^3} + d\sqrt{63t^3}$

20. $x\sqrt{80b^2} + y\sqrt{245b^2}$

21. $8x\sqrt{176x^2} + 7y\sqrt{275x^2}$

22. $10r\sqrt{722t^2} - 3s\sqrt{242t^2}$

23. $3c\sqrt{245d^2} - 7d\sqrt{405d^2}$

24. $2t\sqrt{96t^2} + 7z\sqrt{150t^2}$

25. $x^2\sqrt{8} - 4x\sqrt{18} + \sqrt{50}$

26. $x^2\sqrt{27} - 3x\sqrt{12} + \sqrt{75}$

27. $x^2\sqrt{27} - x\sqrt{147} - \sqrt{12}$

28. $x^2\sqrt{75} - x\sqrt{27} + \sqrt{3}$

29. $x^2\sqrt{3} - y^2\sqrt{3}$

30. $4a^2\sqrt{2} - 9b^2\sqrt{2}$

31. $x^2\sqrt{5} + 2xy\sqrt{5} + y^2\sqrt{5}$

32. $x^2\sqrt{7} + 3xy\sqrt{7} + 2y^2\sqrt{7}$

33. $x^2\sqrt{3} + xy\sqrt{75} + y^2\sqrt{108}$

34. $y^2\sqrt{32} + xy\sqrt{32} + \sqrt{2x^2}$

Developing Skills in Algebra   Book D

Name _____

Date _____ Period _____

Factor each expression. Radical expressions should be in simplified form. Assume all variables represent non-negative numbers.

1. $5a\sqrt{2} + 7b\sqrt{2}$ $\sqrt{2}(5a + 7b)$

2. $5x\sqrt{3} - 10y\sqrt{3}$

3. $4m\sqrt{5} - 3n\sqrt{5}$

4. $2r\sqrt{7} + 9s\sqrt{7}$

5. $5x\sqrt{3x} + 2y\sqrt{3x}$

6. $8a\sqrt{2b} - 4b\sqrt{2b}$

7. $3a\sqrt{7b} - 6b\sqrt{7b}$

8. $10m\sqrt{3m} - 12n\sqrt{3m}$

9. $x^2\sqrt{11} + 8x\sqrt{11} + 2\sqrt{11}$

10. $x^2\sqrt{7} - 7x\sqrt{7} - 3\sqrt{7}$

11. $x^2\sqrt{5} + 5x\sqrt{5} + 10\sqrt{5}$

12. $x^2\sqrt{13} - 3x\sqrt{13} + \sqrt{13}$

13. $a\sqrt{50} + 5\sqrt{162}$

14. $c\sqrt{243} - d\sqrt{48}$

15. $t\sqrt{80} - s\sqrt{405}$

16. $x\sqrt{252} + y\sqrt{847}$

17. $m\sqrt{75n^2} - n\sqrt{48n^2}$

18. $r\sqrt{450t^2} - t\sqrt{800t^2}$

19. $p\sqrt{507q^3} + q\sqrt{300q}$

20. $c\sqrt{128d} + d\sqrt{882d^3}$

21. $5a\sqrt{396a^2} - 10b\sqrt{539a^2}$

22. $6m\sqrt{252m^2} - 3n\sqrt{175m^2}$

23. $14t\sqrt{75t^2} - 7v\sqrt{108t^2}$

24. $8x\sqrt{425y^2} - 16y\sqrt{612y^2}$

25. $x^2\sqrt{12} - 3x\sqrt{27} + \sqrt{243}$

26. $x^2\sqrt{72} - 2x\sqrt{32} - \sqrt{98}$

27. $x^2\sqrt{80} - 4x\sqrt{125} - \sqrt{405}$

28. $x^2\sqrt{112} - 7x\sqrt{252} + \sqrt{567}$

29. $a^2\sqrt{11} - b^2\sqrt{11}$

30. $9a^2\sqrt{3} - 36b^2\sqrt{3}$

31. $a^2\sqrt{7} + 5a\sqrt{7} - 6\sqrt{7}$

32. $x^2\sqrt{5} + 7x\sqrt{5} + 12\sqrt{5}$

33. $a^2\sqrt{3} - a\sqrt{147} + \sqrt{108}$

34. $a^2\sqrt{2} - a\sqrt{18} - \sqrt{32}$

Developing Skills in Algebra   Book D

Addition and Subtraction of Radical Expressions

Name _____

Date _____ Period _____

Simplify.

1. $\sqrt{2} + 3\sqrt{2}$ $4\sqrt{2}$

2. $5\sqrt{3} + 2\sqrt{3}$

3. $2\sqrt{7} - 5\sqrt{7}$

4. $8\sqrt{5} - 3\sqrt{5}$

5. $3\sqrt{5} + 2\sqrt{5} - 3\sqrt{5}$

6. $2\sqrt{11} - 3\sqrt{11} - 6\sqrt{11}$

7. $3\sqrt{13} - 5\sqrt{13} + 2\sqrt{13}$

8. $2\sqrt{15} + 3\sqrt{15} + 2\sqrt{15}$

9. $3\sqrt{3} + 4\sqrt{3} - 9\sqrt{3}$

10. $6\sqrt{5} + \sqrt{5} - 7\sqrt{5}$

11. $8\sqrt{7} + 3\sqrt{7} - 5\sqrt{7}$

12. $6\sqrt{11} - 5\sqrt{11} - 9\sqrt{11}$

13. $5\sqrt{3} - 6\sqrt{2} + 2\sqrt{3}$

14. $8\sqrt{2} + 7\sqrt{3} - 9\sqrt{2}$

15. $3\sqrt{10} - 2\sqrt{5} + 6\sqrt{10}$

16. $4\sqrt{6} - 3\sqrt{2} - 7\sqrt{6}$

17. $2\sqrt{8} + 3\sqrt{2} - 5\sqrt{18}$

18. $5\sqrt{27} - 4\sqrt{3} + 2\sqrt{12}$

19. $4\sqrt{5} - 3\sqrt{125} + 4\sqrt{80}$

20. $2\sqrt{112} + 3\sqrt{63} - 8\sqrt{175}$

21. $\sqrt{\dfrac{3}{8}} - \sqrt{\dfrac{6}{2}} + \sqrt{\dfrac{2}{27}}$

22. $\sqrt{\dfrac{7}{18}} + \sqrt{\dfrac{5}{8}} - \sqrt{\dfrac{7}{2}}$

23. $\sqrt{\dfrac{15}{9}} - \sqrt{\dfrac{7}{12}} - \sqrt{\dfrac{5}{3}}$

24. $\sqrt{\dfrac{3}{5}} - \sqrt{\dfrac{7}{20}} + \sqrt{\dfrac{3}{125}}$

25. $\sqrt{\dfrac{5}{6}} + \sqrt{\dfrac{7}{12}} + \sqrt{\dfrac{6}{5}}$

26. $\sqrt{\dfrac{10}{49}} + \sqrt{\dfrac{5}{2}} + \sqrt{\dfrac{2}{5}}$

27. $\sqrt{\dfrac{14}{169}} - \sqrt{\dfrac{7}{2}} + \sqrt{\dfrac{2}{7}}$

28. $\sqrt{\dfrac{15}{16}} - \sqrt{\dfrac{3}{5}} - \sqrt{\dfrac{5}{3}}$

31

Addition and Subtraction of Radical Expressions

Simplify.

1. $3\sqrt{5} - 7\sqrt{5}$ $-4\sqrt{5}$

2. $2\sqrt{7} + 8\sqrt{7}$

3. $6\sqrt{3} + 10\sqrt{3}$

4. $14\sqrt{2} - 8\sqrt{2}$

5. $6\sqrt{2} - 7\sqrt{2} + 8\sqrt{2}$

6. $14\sqrt{5} - 3\sqrt{5} - 11\sqrt{5}$

7. $8\sqrt{15} - 3\sqrt{15} - 7\sqrt{15}$

8. $13\sqrt{7} - 6\sqrt{7} + 2\sqrt{7}$

9. $5\sqrt{13} - 9\sqrt{13} + 14\sqrt{13}$

10. $6\sqrt{3} - 2\sqrt{3} + 8\sqrt{3}$

11. $6\sqrt{17} - 11\sqrt{17} + 5\sqrt{17}$

12. $8\sqrt{10} + 3\sqrt{10} - 9\sqrt{10}$

13. $3\sqrt{2} - 6\sqrt{3} - 8\sqrt{3}$

14. $4\sqrt{7} - 4\sqrt{10} + 8\sqrt{7}$

15. $14\sqrt{19} + 5\sqrt{13} - 21\sqrt{19}$

16. $15\sqrt{21} + 3\sqrt{17} + 16\sqrt{21}$

17. $9\sqrt{48} - 3\sqrt{27} + 12\sqrt{147}$

18. $15\sqrt{18} - 3\sqrt{50} - 8\sqrt{98}$

19. $5\sqrt{11} + 3\sqrt{484} - 10\sqrt{99}$

20. $13\sqrt{90} - 4\sqrt{40} + 7\sqrt{250}$

21. $\sqrt{\dfrac{80}{3}} + \sqrt{\dfrac{15}{16}} + \sqrt{\dfrac{27}{5}}$

22. $\sqrt{\dfrac{8}{5}} - \sqrt{\dfrac{40}{49}} - \sqrt{\dfrac{45}{2}}$

23. $\sqrt{\dfrac{96}{25}} - \sqrt{\dfrac{98}{3}} + \sqrt{\dfrac{75}{2}}$

24. $\sqrt{\dfrac{25}{14}} + \sqrt{\dfrac{63}{2}} - \sqrt{\dfrac{50}{7}}$

25. $\sqrt{\dfrac{112}{3}} - \sqrt{\dfrac{75}{7}} - \sqrt{\dfrac{84}{25}}$

26. $\sqrt{\dfrac{4}{15}} + \sqrt{\dfrac{48}{10}} - \sqrt{\dfrac{245}{6}}$

27. $\sqrt{\dfrac{25}{6}} + \sqrt{\dfrac{75}{2}} + \sqrt{\dfrac{50}{3}}$

28. $\sqrt{\dfrac{98}{5}} - \sqrt{\dfrac{80}{2}} + \sqrt{\dfrac{49}{10}}$

Addition and Subtraction of Radical Expressions

Name _____

Date _____ Period _____

Simplify.

1. $5a\sqrt{99b^2} + 2b\sqrt{275a^2}$ $25ab\sqrt{11}$

2. $10c\sqrt{6d^2} - 13d\sqrt{294c^2}$

3. $10c\sqrt{112d^2} - 14d\sqrt{343c^2}$

4. $12r\sqrt{98s^2} + 15s\sqrt{162r^2}$

5. $3rs\sqrt{325} - 10\sqrt{832r^2s^2}$

6. $15\sqrt{243m^2n^2} - 29mn\sqrt{108}$

7. $18a\sqrt{490b^2} + 13\sqrt{1210a^2b^2}$

8. $27\sqrt{245r^2s^2} - 23rs\sqrt{845}$

9. $26a\sqrt{363a^3} + 15a^2\sqrt{507a}$

10. $16ab\sqrt{847b} + 13\sqrt{175a^2b^3}$

11. $26yz\sqrt{486z} - 19\sqrt{726y^2z^3}$

12. $23\sqrt{405x^3y^2} - 12xy\sqrt{980x}$

13. $15a^2b\sqrt{250b^3} + 13\sqrt{810a^4b^5}$

14. $10cd^2\sqrt{468c^3} - 15c^2d^2\sqrt{1300c}$

15. $13mn\sqrt{288m^3n^2} - 15\sqrt{392m^5n^4}$

16. $25rs\sqrt{396r^3s^2} + 14\sqrt{891r^5s^4}$

17. $23\sqrt{48x^3y^3} + 10xy\sqrt{75xy}$

18. $13st\sqrt{567s^3t^3} - 22s^2t^2\sqrt{1008st}$

19. $13\sqrt{640m^5n^5} - 12m^2n^2\sqrt{1440mn}$

20. $19yz^2\sqrt{605y^3z} + 17z\sqrt{320y^5z^3}$

21. $\sqrt{\dfrac{180b^2}{7a}} + \sqrt{\dfrac{45b^2}{7a}}$

22. $\sqrt{\dfrac{448r^2}{5s}} - \sqrt{\dfrac{700r^2}{5s}}$

23. $\sqrt{\dfrac{160x^2}{3y}} - \sqrt{\dfrac{1000x^2}{3y}}$

24. $\sqrt{\dfrac{338a^4}{11b}} + \sqrt{\dfrac{450a^4}{11b}}$

25. $\sqrt{\dfrac{208a}{5b}} - \sqrt{\dfrac{125a}{13b}}$

26. $\sqrt{\dfrac{192x^2}{7y^3}} - \sqrt{\dfrac{343x^2}{3y^3}}$

27. $\sqrt{\dfrac{150a^3}{11b}} + \sqrt{\dfrac{176a^3}{6b}}$

28. $\sqrt{\dfrac{539x^2}{2y^3}} - \sqrt{\dfrac{50x^2}{11y^3}}$

33

Addition and Subtraction of Radical Expressions   Name _____

Date _____ Period _____

Simplify.

1. $10x\sqrt{1690y^2} - 12y\sqrt{360x^2}$ $58\,xy\sqrt{10}$

2. $15ab\sqrt{252} - 7\sqrt{1183a^2b^2}$

3. $15c\sqrt{80d^2} + 12d\sqrt{125c^2}$

4. $16r\sqrt{216s^2} - 20\sqrt{384r^2s^2}$

5. $11pq\sqrt{117} - 13\sqrt{637p^2q^2}$

6. $21mn\sqrt{72} + 12\sqrt{200m^2n^2}$

7. $13v\sqrt{147w^2} + 12\sqrt{300v^2w^2}$

8. $8pq\sqrt{1100} + 12\sqrt{1331p^2q^2}$

9. $12a\sqrt{500a^2b} - 13a^2\sqrt{720b}$

10. $14xy\sqrt{128x^3} - 17\sqrt{128x^5y^2}$

11. $21cd\sqrt{539c^3} + 14c^2\sqrt{704cd^2}$

12. $23b^2c\sqrt{490c^3} - 15bc^2\sqrt{250b^2c}$

13. $17x^2y\sqrt{108y^3} - 21xy^2\sqrt{192x^2y}$

14. $11x^2y\sqrt{325xy^2} + 21xy\sqrt{1053x^3y^2}$

15. $31cd\sqrt{54c^2d^3} - 15c^2\sqrt{600d^5}$

16. $29ab\sqrt{63a^2b} - 19a\sqrt{448a^2b^3}$

17. $25c^2d^2\sqrt{180cd} + 17\sqrt{20c^5d^5}$

18. $14x^2y^2\sqrt{1100xy} - 27xy\sqrt{704x^3y^3}$

19. $17ab\sqrt{468a^3b^3} + 24\sqrt{832a^5b^5}$

20. $30pq\sqrt{75p^3q^3} + 15p^2q^2\sqrt{243pq}$

21. $\sqrt{\dfrac{200a^2}{11b}} - \sqrt{\dfrac{275a^2}{2b}}$

22. $\sqrt{\dfrac{150x}{13y}} + \sqrt{\dfrac{637x}{6y}}$

23. $\sqrt{\dfrac{640p^2}{3q}} + \sqrt{\dfrac{147p^2}{10q}}$

24. $\sqrt{\dfrac{245m}{2n}} - \sqrt{\dfrac{162m}{5n}}$

25. $\sqrt{\dfrac{112r}{5s}} - \sqrt{\dfrac{320r}{7s}}$

26. $\sqrt{\dfrac{300x^2}{7y}} - \sqrt{\dfrac{112x^2}{3y}}$

27. $\sqrt{\dfrac{117v^2}{10w}} + \sqrt{\dfrac{90v^2}{13w}}$

28. $\sqrt{\dfrac{275a^2}{6b}} + \sqrt{\dfrac{294a^2}{11b}}$

34   Developing Skills in Algebra   Book D

Multiplication of Binomials Containing Radicals

Name _____

Date _____ Period _____

Multiply. Radical expressions should be in simplified form.

1. $\left(\sqrt{3} + \sqrt{2}\right)\left(\sqrt{3} + \sqrt{5}\right)$ $3 + \sqrt{6} + \sqrt{15} + \sqrt{10}$

2. $\left(\sqrt{5} + \sqrt{7}\right)\left(\sqrt{14} - \sqrt{8}\right)$

3. $\left(\sqrt{2} + \sqrt{7}\right)\left(\sqrt{11} - \sqrt{5}\right)$

4. $\left(\sqrt{6} + \sqrt{2}\right)\left(4\sqrt{3} + \sqrt{2}\right)$

5. $\left(\sqrt{8} - \sqrt{5}\right)\left(3\sqrt{13} + \sqrt{8}\right)$

6. $\left(\sqrt{10} + \sqrt{2}\right)\left(\sqrt{5} - \sqrt{3}\right)$

7. $\left(\sqrt{14} + \sqrt{5}\right)\left(\sqrt{14} - \sqrt{5}\right)$

8. $\left(2\sqrt{5} + \sqrt{2}\right)\left(\sqrt{13} - \sqrt{5}\right)$

9. $\left(\sqrt{6} - \sqrt{5}\right)\left(\sqrt{3} + \sqrt{8}\right)$

10. $\left(\sqrt{7} + \sqrt{5}\right)\left(3\sqrt{2} + \sqrt{5}\right)$

11. $\left(\sqrt{3} + \sqrt{6}\right)\left(\sqrt{5} - \sqrt{2}\right)$

12. $\left(\sqrt{14} + \sqrt{7}\right)\left(7\sqrt{7} - \sqrt{5}\right)$

13. $\left(\sqrt{5} + \sqrt{3}\right)\left(5\sqrt{2} - \sqrt{6}\right)$

14. $\left(\sqrt{10} - \sqrt{7}\right)\left(2\sqrt{15} + \sqrt{3}\right)$

15. $\left(\sqrt{11} - \sqrt{7}\right)\left(\sqrt{8} - \sqrt{5}\right)$

16. $\left(7\sqrt{12} + \sqrt{2}\right)\left(\sqrt{10} + \sqrt{8}\right)$

17. $\left(\sqrt{13} + \sqrt{3}\right)\left(\sqrt{13} - \sqrt{7}\right)$

18. $\left(\sqrt{13} + \sqrt{7}\right)\left(\sqrt{3} - \sqrt{6}\right)$

19. $\left(\sqrt{2} + \sqrt{5}\right)\left(7\sqrt{11} - \sqrt{8}\right)$

20. $\left(\sqrt{12} - \sqrt{8}\right)\left(\sqrt{12} + \sqrt{8}\right)$

21. $\left(2\sqrt{7} + \sqrt{3}\right)^2$

22. $\left(\sqrt{5} + \sqrt{3}\right)^2$

23. $\left(\sqrt{8} - \sqrt{5}\right)^2$

24. $\left(3\sqrt{3} + \sqrt{8}\right)^2$

25. $\left(\sqrt{12} + \sqrt{5}\right)^2$

26. $\left(5\sqrt{6} + \sqrt{3}\right)^2$

27. $\left(\sqrt{11} - \sqrt{5}\right)^2$

28. $\left(\sqrt{7} + \sqrt{5}\right)^2$

29. $\left(\sqrt{7} - \sqrt{5}\right)\left(\sqrt{7} + \sqrt{5}\right)$

30. $\left(\sqrt{12} - \sqrt{5}\right)\left(\sqrt{12} + \sqrt{5}\right)$

31. $\left(\sqrt{8} + \sqrt{3}\right)\left(\sqrt{8} - \sqrt{3}\right)$

32. $\left(\sqrt{15} - \sqrt{2}\right)\left(\sqrt{15} + \sqrt{2}\right)$

33. $\left(\sqrt{10} - \sqrt{6}\right)\left(\sqrt{10} + \sqrt{6}\right)$

34. $\left(\sqrt{27} - \sqrt{6}\right)\left(\sqrt{27} + \sqrt{6}\right)$

Multiplication of Binomials Containing Radicals

Name _____

Date _____ Period _____

Multiply. Radical expressions should be in simplified form.

1. $\left(\sqrt{5} + \sqrt{3}\right)\left(\sqrt{2} - \sqrt{3}\right)$ $\sqrt{10} - \sqrt{15} + \sqrt{6} - 3$

2. $\left(\sqrt{8} + \sqrt{6}\right)\left(\sqrt{10} - \sqrt{2}\right)$

3. $\left(8\sqrt{2} - \sqrt{5}\right)\left(\sqrt{6} + \sqrt{5}\right)$

4. $\left(\sqrt{11} + 3\sqrt{6}\right)\left(\sqrt{14} - \sqrt{7}\right)$

5. $\left(\sqrt{14} + \sqrt{3}\right)\left(\sqrt{14} - \sqrt{2}\right)$

6. $\left(\sqrt{8} + \sqrt{7}\right)\left(2\sqrt{3} + \sqrt{5}\right)$

7. $\left(3\sqrt{3} - \sqrt{5}\right)\left(\sqrt{6} + \sqrt{3}\right)$

8. $\left(\sqrt{13} + 5\sqrt{2}\right)\left(\sqrt{11} + \sqrt{6}\right)$

9. $\left(2\sqrt{6} + \sqrt{7}\right)\left(\sqrt{7} - \sqrt{3}\right)$

10. $\left(\sqrt{13} - \sqrt{8}\right)\left(2\sqrt{5} + \sqrt{6}\right)$

11. $\left(\sqrt{10} + 2\sqrt{3}\right)\left(\sqrt{12} + \sqrt{7}\right)$

12. $\left(\sqrt{15} + \sqrt{3}\right)\left(\sqrt{8} - 3\sqrt{3}\right)$

13. $\left(\sqrt{11} - 3\sqrt{3}\right)\left(\sqrt{13} + \sqrt{2}\right)$

14. $\left(\sqrt{11} + \sqrt{5}\right)\left(\sqrt{3} - 4\sqrt{7}\right)$

15. $\left(\sqrt{12} + \sqrt{6}\right)\left(\sqrt{7} - \sqrt{2}\right)$

16. $\left(\sqrt{2} - \sqrt{7}\right)\left(\sqrt{2} + \sqrt{7}\right)$

17. $\left(\sqrt{6} + \sqrt{3}\right)\left(2\sqrt{10} + \sqrt{7}\right)$

18. $\left(\sqrt{11} + \sqrt{8}\right)\left(2\sqrt{8} - \sqrt{6}\right)$

19. $\left(\sqrt{2} - 5\sqrt{3}\right)\left(\sqrt{8} + \sqrt{7}\right)$

20. $\left(\sqrt{6} - 2\sqrt{7}\right)\left(\sqrt{13} - \sqrt{6}\right)$

21. $\left(\sqrt{10} + \sqrt{2}\right)^2$

22. $\left(\sqrt{12} + \sqrt{5}\right)^2$

23. $\left(\sqrt{11} - \sqrt{3}\right)^2$

24. $\left(\sqrt{10} - \sqrt{5}\right)^2$

25. $\left(\sqrt{12} + 3\sqrt{6}\right)^2$

26. $\left(\sqrt{20} + 2\sqrt{3}\right)^2$

27. $\left(\sqrt{20} + \sqrt{2}\right)^2$

28. $\left(\sqrt{30} - \sqrt{7}\right)^2$

29. $\left(\sqrt{6} - \sqrt{13}\right)\left(\sqrt{6} + \sqrt{13}\right)$

30. $\left(\sqrt{11} + \sqrt{3}\right)\left(\sqrt{11} - \sqrt{3}\right)$

31. $\left(\sqrt{12} - \sqrt{7}\right)\left(\sqrt{12} + \sqrt{7}\right)$

32. $\left(\sqrt{5} - 3\sqrt{2}\right)\left(\sqrt{5} + 3\sqrt{2}\right)$

33. $\left(\sqrt{15} + \sqrt{2}\right)\left(\sqrt{15} - \sqrt{2}\right)$

34. $\left(\sqrt{13} - \sqrt{5}\right)\left(\sqrt{13} + \sqrt{5}\right)$

Developing Skills in Algebra   Book D

Multiplication of Binomials Containing Radicals   Name _____

Date _____ Period _____

Multiply. Radical expressions should be in simplified form.

1. $(\sqrt{11} + \sqrt{3})(\sqrt{12} - \sqrt{5})$ $\sqrt{132} - \sqrt{55} + 6 - \sqrt{15}$ **2.** $(\sqrt{17} - \sqrt{2})(\sqrt{6} + \sqrt{2})$

3. $(\sqrt{3} + \sqrt{5})(\sqrt{8} - \sqrt{3})$ **4.** $(\sqrt{2} - \sqrt{6})(\sqrt{11} - \sqrt{3})$

5. $(\sqrt{14} - 3\sqrt{2})(\sqrt{6} - 5\sqrt{2})$ **6.** $(\sqrt{5} + 2\sqrt{6})(\sqrt{5} - \sqrt{7})$

7. $(\sqrt{10} + \sqrt{5})(\sqrt{3} + 2\sqrt{7})$ **8.** $(\sqrt{7} - 3\sqrt{3})(\sqrt{10} + \sqrt{3})$

9. $(\sqrt{7} - \sqrt{2})(\sqrt{7} + 3\sqrt{2})$ **10.** $(\sqrt{11} - \sqrt{2})(\sqrt{13} - \sqrt{6})$

11. $(\sqrt{8} - 3\sqrt{5})(\sqrt{7} - \sqrt{8})$ **12.** $(\sqrt{7} + \sqrt{6})(\sqrt{10} - \sqrt{5})$

13. $(\sqrt{7} + \sqrt{3})(\sqrt{11} - 2\sqrt{7})$ **14.** $(\sqrt{10} + \sqrt{8})(3\sqrt{6} - \sqrt{7})$

15. $(\sqrt{12} + \sqrt{7})(\sqrt{12} - \sqrt{11})$ **16.** $(\sqrt{13} + \sqrt{3})(\sqrt{14} - 2\sqrt{6})$

17. $(\sqrt{14} - \sqrt{8})(\sqrt{13} - 2\sqrt{7})$ **18.** $(\sqrt{13} - \sqrt{5})(\sqrt{6} - \sqrt{8})$

19. $(\sqrt{12} + \sqrt{5})(\sqrt{12} + \sqrt{2})$ **20.** $(\sqrt{15} - \sqrt{2})(\sqrt{14} + 3\sqrt{3})$

21. $(\sqrt{13} + \sqrt{5})^2$ **22.** $(\sqrt{11} - \sqrt{6})^2$

23. $(\sqrt{18} - \sqrt{3})^2$ **24.** $(\sqrt{20} + \sqrt{6})^2$

25. $(\sqrt{11} + \sqrt{5})^2$ **26.** $(\sqrt{18} - \sqrt{5})^2$

27. $(\sqrt{12} - \sqrt{7})^2$ **28.** $(\sqrt{30} + \sqrt{2})^2$

29. $(\sqrt{6} + \sqrt{8})(\sqrt{6} - \sqrt{8})$ **30.** $(\sqrt{15} + \sqrt{7})(\sqrt{15} - \sqrt{7})$

31. $(\sqrt{11} - \sqrt{5})(\sqrt{11} + \sqrt{5})$ **32.** $(\sqrt{7} - \sqrt{11})(\sqrt{7} + \sqrt{11})$

33. $(\sqrt{12} + \sqrt{6})(\sqrt{12} - \sqrt{6})$ **34.** $(\sqrt{17} - \sqrt{5})(\sqrt{17} + \sqrt{5})$

37

Multiplication of Binomials Containing Radicals

Multiply. Radical expressions should be in simplified form.

1. $\left(\sqrt{7} + \sqrt{8}\right)\left(\sqrt{3} - \sqrt{2}\right)$ $\sqrt{21} - \sqrt{14} + 2\sqrt{6} - 4$

2. $\left(\sqrt{5} + 2\sqrt{6}\right)\left(\sqrt{6} - \sqrt{2}\right)$

3. $\left(2\sqrt{3} - \sqrt{5}\right)\left(\sqrt{5} + \sqrt{8}\right)$

4. $\left(\sqrt{3} + \sqrt{5}\right)\left(\sqrt{7} - \sqrt{8}\right)$

5. $\left(2\sqrt{2} + \sqrt{7}\right)\left(\sqrt{8} - \sqrt{7}\right)$

6. $\left(\sqrt{6} - 2\sqrt{7}\right)\left(\sqrt{5} - \sqrt{7}\right)$

7. $\left(\sqrt{5} - \sqrt{8}\right)\left(\sqrt{11} + \sqrt{2}\right)$

8. $\left(\sqrt{8} + \sqrt{3}\right)\left(\sqrt{10} + \sqrt{2}\right)$

9. $\left(\sqrt{13} + 2\sqrt{6}\right)\left(\sqrt{12} + \sqrt{6}\right)$

10. $\left(2\sqrt{3} - \sqrt{7}\right)\left(\sqrt{6} - \sqrt{3}\right)$

11. $\left(\sqrt{10} - \sqrt{6}\right)\left(\sqrt{13} - \sqrt{3}\right)$

12. $\left(\sqrt{11} + \sqrt{3}\right)\left(\sqrt{8} - \sqrt{3}\right)$

13. $\left(\sqrt{3} + \sqrt{8}\right)\left(\sqrt{12} - \sqrt{5}\right)$

14. $\left(\sqrt{6} - \sqrt{8}\right)\left(\sqrt{7} - \sqrt{2}\right)$

15. $\left(\sqrt{8} - 2\sqrt{3}\right)\left(\sqrt{7} + \sqrt{3}\right)$

16. $\left(\sqrt{12} + \sqrt{5}\right)\left(\sqrt{8} + \sqrt{5}\right)$

17. $\left(\sqrt{6} + \sqrt{8}\right)\left(\sqrt{10} - \sqrt{6}\right)$

18. $\left(\sqrt{8} - \sqrt{7}\right)\left(\sqrt{7} - \sqrt{3}\right)$

19. $\left(\sqrt{3} - \sqrt{7}\right)\left(3\sqrt{7} + \sqrt{6}\right)$

20. $\left(\sqrt{7} + \sqrt{3}\right)\left(\sqrt{8} - \sqrt{6}\right)$

21. $\left(\sqrt{13} + \sqrt{2}\right)^2$

22. $\left(\sqrt{5} - 5\sqrt{3}\right)^2$

23. $\left(\sqrt{11} - 2\sqrt{7}\right)^2$

24. $\left(\sqrt{8} + \sqrt{5}\right)^2$

25. $\left(\sqrt{10} + \sqrt{6}\right)^2$

26. $\left(\sqrt{6} + 2\sqrt{7}\right)^2$

27. $\left(\sqrt{30} - \sqrt{5}\right)^2$

28. $\left(\sqrt{14} - \sqrt{5}\right)^2$

29. $\left(\sqrt{5} + 2\sqrt{3}\right)\left(\sqrt{5} - 2\sqrt{3}\right)$

30. $\left(\sqrt{8} + \sqrt{13}\right)\left(\sqrt{8} - \sqrt{13}\right)$

31. $\left(\sqrt{20} - \sqrt{7}\right)\left(\sqrt{20} + \sqrt{7}\right)$

32. $\left(\sqrt{6} - \sqrt{7}\right)\left(\sqrt{6} + \sqrt{7}\right)$

33. $\left(\sqrt{15} + \sqrt{12}\right)\left(\sqrt{15} - \sqrt{12}\right)$

34. $\left(\sqrt{13} + \sqrt{2}\right)\left(\sqrt{13} - \sqrt{2}\right)$

Fractions Containing Radicals

Simplify.

1. $\dfrac{1}{\sqrt{2} + \sqrt{3}}$ $\sqrt{3} - \sqrt{2}$

2. $\dfrac{2}{\sqrt{5} + \sqrt{8}}$

3. $\dfrac{4}{\sqrt{10} + \sqrt{2}}$

4. $\dfrac{6}{\sqrt{2} - \sqrt{5}}$

5. $\dfrac{3}{\sqrt{3} - \sqrt{5}}$

6. $\dfrac{9}{\sqrt{5} - \sqrt{6}}$

7. $\dfrac{4}{\sqrt{2} - \sqrt{6}}$

8. $\dfrac{12}{\sqrt{12} - \sqrt{2}}$

9. $\dfrac{29}{\sqrt{31} - \sqrt{9}}$

10. $\dfrac{5}{\sqrt{13} - \sqrt{2}}$

11. $\dfrac{17}{\sqrt{14} + 3\sqrt{2}}$

12. $\dfrac{15}{\sqrt{17} + \sqrt{11}}$

13. $\dfrac{15}{\sqrt{6} - \sqrt{7}}$

14. $\dfrac{8}{\sqrt{17} - \sqrt{3}}$

15. $\dfrac{7}{\sqrt{3} - \sqrt{7}}$

16. $\dfrac{10}{\sqrt{18} - 2\sqrt{3}}$

17. $\dfrac{16}{\sqrt{16} + 2\sqrt{5}}$

18. $\dfrac{14}{\sqrt{19} + \sqrt{3}}$

19. $\dfrac{10}{\sqrt{13} - \sqrt{5}}$

20. $\dfrac{18}{\sqrt{17} - \sqrt{3}}$

21. $\dfrac{20}{\sqrt{12} - \sqrt{6}}$

22. $\dfrac{23}{\sqrt{18} - \sqrt{5}}$

23. $\dfrac{25}{\sqrt{14} + \sqrt{5}}$

24. $\dfrac{27}{\sqrt{20} - \sqrt{3}}$

25. $\dfrac{26}{\sqrt{19} + \sqrt{2}}$

26. $\dfrac{22}{\sqrt{31} - 2\sqrt{2}}$

Fractions Containing Radicals

Name _____

Date _____ Period _____

Simplify.

1. $\dfrac{14}{\sqrt{12} - \sqrt{5}}$ $4\sqrt{3} \;+\; 2\sqrt{5}$

2. $\dfrac{4}{\sqrt{2} + \sqrt{6}}$

3. $\dfrac{1}{\sqrt{8} - \sqrt{3}}$

4. $\dfrac{13}{\sqrt{12} + 5}$

5. $\dfrac{5}{\sqrt{10} + \sqrt{3}}$

6. $\dfrac{17}{\sqrt{2} + \sqrt{7}}$

7. $\dfrac{18}{\sqrt{14} - \sqrt{2}}$

8. $\dfrac{2}{\sqrt{3} - \sqrt{6}}$

9. $\dfrac{7}{\sqrt{15} + 3}$

10. $\dfrac{11}{\sqrt{7} + 8}$

11. $\dfrac{9}{\sqrt{14} + 2\sqrt{3}}$

12. $\dfrac{3}{2\sqrt{5} - \sqrt{7}}$

13. $\dfrac{5}{\sqrt{15} + \sqrt{2}}$

14. $\dfrac{22}{\sqrt{10} - 5}$

15. $\dfrac{24}{\sqrt{20} + 2}$

16. $\dfrac{6}{\sqrt{12} + \sqrt{6}}$

17. $\dfrac{12}{\sqrt{7} - 4}$

18. $\dfrac{23}{3 - \sqrt{8}}$

19. $\dfrac{18}{\sqrt{14} + 2\sqrt{6}}$

20. $\dfrac{10}{2\sqrt{6} + \sqrt{8}}$

21. $\dfrac{24}{\sqrt{7} + 4}$

22. $\dfrac{19}{13 + \sqrt{3}}$

23. $\dfrac{15}{\sqrt{12} - \sqrt{5}}$

24. $\dfrac{25}{\sqrt{17} - \sqrt{5}}$

25. $\dfrac{27}{\sqrt{7} - \sqrt{13}}$

26. $\dfrac{21}{\sqrt{14} - \sqrt{7}}$

Simple Radical Equations

Name _____

Date _____ Period _____

Solve. Solutions must be checked.

1. $\sqrt{x} = 7$ $x = 49$

2. $\sqrt{x} = 3$

3. $\sqrt{x} = 15$

4. $\sqrt{x} = 13$

5. $\sqrt{x} = 2\sqrt{3}$

6. $\sqrt{x} = 5\sqrt{2}$

7. $\sqrt{x} = 7\sqrt{6}$

8. $\sqrt{x} = 3\sqrt{10}$

9. $\sqrt{x + 27} = 5$

10. $\sqrt{x - 4} = 3$

11. $\sqrt{x - 6} = 11$

12. $\sqrt{x + 13} = 3$

13. $\sqrt{2x + 3} = 5$

14. $\sqrt{10 - 3x} = 7$

15. $\sqrt{16 - 5x} = 6$

16. $\sqrt{4x + 13} = 2$

17. $\sqrt{3x - 6} = 2\sqrt{3}$

18. $\sqrt{2 - 5x} = 4\sqrt{7}$

19. $\sqrt{2x + 13} = \sqrt{3}$

20. $\sqrt{7x + 4} = 5\sqrt{8}$

21. $\sqrt{x} = \sqrt{3} - 2$

22. $\sqrt{x} = \sqrt{3} - \sqrt{5}$

23. $\sqrt{x} = 7 - 2\sqrt{5}$

24. $\sqrt{x} = \sqrt{11} + 2\sqrt{2}$

25. $\sqrt{x + 3} = \sqrt{3} - \sqrt{7}$

26. $\sqrt{x - 4} = \sqrt{11} + 3\sqrt{5}$

27. $\sqrt{x + 7} = 2\sqrt{3} + 3\sqrt{7}$

28. $\sqrt{x - 8} = 4\sqrt{6} - 3\sqrt{10}$

29. $\sqrt{3x - 2} = \sqrt{6} - \sqrt{5}$

30. $\sqrt{7x + 1} = \sqrt{5} + 3$

31. $\sqrt{5x + 1} = 3\sqrt{2} - \sqrt{7}$

32. $\sqrt{2x + 7} = 3\sqrt{3} - \sqrt{5}$

33. $\sqrt{4x - 6} = \sqrt{3} + 4$

34. $\sqrt{9x + 2} = \sqrt{7} - \sqrt{11}$

Simple Radical Equations

Solve. Solutions must be checked.

1. $\sqrt{x} = 10$ $x = 100$

2. $\sqrt{x} = 4$

3. $\sqrt{x} = 25$

4. $\sqrt{x} = 7$

5. $\sqrt{x} = 3\sqrt{5}$

6. $\sqrt{x} = 8\sqrt{2}$

7. $\sqrt{x} = 4\sqrt{11}$

8. $\sqrt{x} = 5\sqrt{12}$

9. $\sqrt{x + 5} = 9$

10. $\sqrt{x + 10} = 3$

11. $\sqrt{x - 7} = 2$

12. $\sqrt{x + 8} = 6$

13. $\sqrt{2x + 5} = 7$

14. $\sqrt{5x - 6} = 3$

15. $\sqrt{7x - 1} = 2\sqrt{5}$

16. $\sqrt{8x + 3} = 9\sqrt{3}$

17. $\sqrt{2 - 4x} = 5\sqrt{2}$

18. $\sqrt{10x - 8} = 4\sqrt{7}$

19. $\sqrt{3x - 8} = 7\sqrt{7}$

20. $\sqrt{9x + 3} = 5\sqrt{3}$

21. $\sqrt{x} = 5 + \sqrt{2}$

22. $\sqrt{x} = \sqrt{2} - \sqrt{7}$

23. $\sqrt{x} = 2\sqrt{10} - \sqrt{3}$

24. $\sqrt{x} = \sqrt{5} + 2\sqrt{10}$

25. $\sqrt{x - 5} = 2\sqrt{7} + \sqrt{3}$

26. $\sqrt{x + 9} = 3\sqrt{5} - 2\sqrt{3}$

27. $\sqrt{x + 2} = \sqrt{3} - \sqrt{10}$

28. $\sqrt{x - 8} = 3\sqrt{6} + 2\sqrt{3}$

29. $\sqrt{3x + 5} = \sqrt{2} - \sqrt{8}$

30. $\sqrt{6 - 7x} = 3\sqrt{5} + 2\sqrt{11}$

31. $\sqrt{4x - 3} = \sqrt{7} + 2\sqrt{2}$

32. $\sqrt{10 - 6x} = \sqrt{9} + 3\sqrt{5}$

33. $\sqrt{7 - 2x} = \sqrt{11} + 2\sqrt{5}$

34. $\sqrt{3x + 4} = \sqrt{3} - \sqrt{7}$

Developing Skills in Algebra   Book D

The Distance Formula

Name _____

Date _____ Period _____

If (x_1, y_1) and (x_2, y_2) are the coordinates of two points in the coordinate plane, then the distance between the points is given by the formula:

$$d = \sqrt{(x_2 - x_1)^2 + (y_2 - y_1)^2}$$

Find the distance between points whose coordinates are given.

1. $(-5, -7); (7, 9)$ 20

2. $(15, 26); (-5, -22)$

3. $(-20, -9); (15, 3)$

4. $(12, 6); (-3, -14)$

5. $(3, -12); (11, 3)$

6. $(18, 35); (-7, -25)$

7. $(14, 3); (7, -21)$

8. $(-10, 16); (5, -20)$

9. $(-6, 15); (14, -6)$

10. $(-12, 3); (3, 11)$

11. $(3, -2); (9, -10)$

12. $(14, 16); (-6, -5)$

13. $(-4, 6); (4, -9)$

14. $(-1, -5); (-11, 19)$

15. $(13, -2); (-7, 19)$

16. $(15, -17); (3, 18)$

17. $(-3, 2); (5, 7)$

18. $(-12, 4); (7, -3)$

19. $(5, -3); (2, 6)$

20. $(6, -17); (18, -5)$

21. $(-8, -10); (7, 3)$

22. $(-13, 12); (15, -3)$

23. $(-3, 12); (11, 4)$

24. $(13, -4); (-5, 12)$

25. $(-8, -2); (-3, 8)$

26. $(15, 23); (2, 5)$

27. $(7, -5); (-10, 3)$

28. $(17, 6); (3, -22)$

29. $(-9, -10); (6, 8)$

30. $(26, 12); (-3, -7)$

The Distance Formula

Name _____

Date _____ Period _____

If (x_1, y_1) and (x_2, y_2) are the coordinates of two points in the coordinate plane, then the distance between the points is given by the formula:

$$d = \sqrt{(x_2 - x_1)^2 + (y_2 - y_1)^2}$$

Find the distance between points whose coordinates are given.

1. (12, 7); (4, −8)   *17*

2. (5, −20); (20, 16)

3. (−2, 10); (−12, −14)

4. (7, 15); (−5, −1)

5. (15, −25); (3, 10)

6. (3, −11); (−12, −3)

7. (−18, −4); (17, 8)

8. (14, 23); (−11, −37)

9. (0, 0); (3, 4)

10. (−10, 5); (10, −16)

11. (−5, 18); (15, −30)

12. (3, −11); (11, 4)

13. (−12, −12); (−5, 12)

14. (2, −13); (−13, 7)

15. (13, 7); (−7, −14)

16. (12, −6); (−6, 18)

17. (−1, 4); (−4, −5)

18. (15, −6); (−4, 12)

19. (16, 3); (12, 10)

20. (23, 5); (−6, −4)

21. (−8, 7); (1, −2)

22. (22, −7); (−3, 12)

23. (3, 13); (5, 7)

24. (−12, −9); (3, 7)

25. (18, −5); (12, 7)

26. (30, −18); (15, −3)

27. (13, 2); (−5, 6)

28. (14, −2); (−3, 21)

29. (28, 15); (−6, −3)

30. (−12, −19); (−5, −13)

Developing Skills in Algebra   Book D

Simple Quadratic Equations

Name _____

Date _____ Period _____

Solve. Express your answers in simplified radical form.

1. $x^2 = 16$ $x = \pm 4$

2. $x^2 = 9$

3. $x^2 = 49$

4. $x^2 = 81$

5. $3x^2 = 108$

6. $5x^2 = 320$

7. $7x^2 = 700$

8. $4x^2 = 484$

9. $6x^2 = 54$

10. $8x^2 = 200$

11. $4x^2 = 2704$

12. $16x^2 = 64$

13. $9x^2 = 2304$

14. $15x^2 = 4335$

15. $4x^2 = 676$

16. $7x^2 = 2268$

17. $15x^2 = 9375$

18. $12x^2 = 2352$

19. $13x^2 = 2925$

20. $5x^2 = 1805$

21. $16x^2 = 25$

22. $49x^2 = 9$

23. $9x^2 = 4$

24. $36x^2 = 121$

25. $25x^2 = 7$

26. $81x^2 = 10$

27. $169x^2 = 14$

28. $225x^2 = 17$

29. $2x^2 = 3$

30. $5x^2 = 7$

31. $7x^2 = 8$

32. $11x^2 = 27$

33. $15x^2 = 176$

34. $20x^2 = 63$

Simple Quadratic Equations

Name _____

Date _____ Period _____

Solve. Express your answers in simplified radical form.

1. $x^2 = 100$ $x = \pm 10$

2. $x^2 = 324$

3. $x^2 = 4$

4. $x^2 = 225$

5. $2x^2 = 50$

6. $13x^2 = 468$

7. $7x^2 = 63$

8. $3x^2 = 243$

9. $9x^2 = 1296$

10. $11x^2 = 2156$

11. $8x^2 = 392$

12. $8x^2 = 128$

13. $5x^2 = 4500$

14. $11x^2 = 1331$

15. $7x^2 = 175$

16. $3x^2 = 768$

17. $10x^2 = 2890$

18. $7x^2 = 448$

19. $2x^2 = 5000$

20. $5x^2 = 845$

21. $36x^2 = 169$

22. $121x^2 = 100$

23. $25x^2 = 49$

24. $225x^2 = 16$

25. $16x^2 = 11$

26. $81x^2 = 33$

27. $64x^2 = 21$

28. $196x^2 = 15$

29. $7x^2 = 11$

30. $6x^2 = 29$

31. $3x^2 = 338$

32. $5x^2 = 147$

33. $10x^2 = 243$

34. $17x^2 = 363$

Simple Quadratic Equations   Name _____

Date _____ Period _____

Solve. Express your answers in simplified radical form.

1. $(x + 2)^2 = 16$ x= 2 or -6   **2.** $(x - 5)^2 = 9$

3. $(x + 11)^2 = 36$ **4.** $(x + 8)^2 = 49$

5. $(x - 12)^2 = 121$ **6.** $(x - 15)^2 = 256$

7. $(x - 9)^2 = 441$ **8.** $(x + 20)^2 = 196$

9. $2(x + 4)^2 = 338$ **10.** $3(x - 13)^2 = 432$

11. $5(x - 2)^2 = 2000$ **12.** $13(x + 6)^2 = 4693$

13. $7(x - 9)^2 = 1575$ **14.** $4(x - 11)^2 = 2704$

15. $12(x + 3)^2 = 7488$ **16.** $6(x + 7)^2 = 1800$

17. $(x + 2)^2 = 75$ **18.** $(x - 5)^2 = 384$

19. $(x - 9)^2 = 588$ **20.** $(x + 6)^2 = 1331$

21. $(x + 7)^2 = 1620$ **22.** $(x + 9)^2 = 252$

23. $(x + 4)^2 = 245$ **24.** $(x - 8)^2 = 1350$

25. $(x - 1)^2 = 864$ **26.** $(x + 5)^2 = 1083$

27. $(x - 4)^2 = 1792$ **28.** $(x - 10)^2 = 891$

29. $3(x + 2)^2 = 100$ **30.** $5(x - 7)^2 = 676$

31. $11(x - 7)^2 = 289$ **32.** $6(x - 1)^2 = 169$

33. $20(x - 3)^2 = 625$ **34.** $27(x + 11)^2 = 900$

Name _____

Date _____ Period _____

Solve. Express your answers in simplified radical form.

1. $(x - 3)^2 = 100$ $x = 7$ or $+13$

2. $(x + 11)^2 = 64$

3. $(x + 8)^2 = 4$

4. $(x - 2)^2 = 256$

5. $(x - 9)^2 = 169$

6. $(x - 13)^2 = 25$

7. $(x + 5)^2 = 81$

8. $(x - 1)^2 = -9$

9. $3(x - 2)^2 = 108$

10. $6(x + 7)^2 = 1350$

11. $5(x + 6)^2 - 80 = 0$

12. $12(x + 1)^2 = 2352$

13. $4(x - 7)^2 = 484$

14. $3(x - 2)^2 - 972 = 0$

15. $9(x - 3)^2 + 441 = 0$

16. $5(x - 9)^2 = 720$

17. $(x + 3)^2 = 567$

18. $(x - 3)^2 = 1176$

19. $(x - 5)^2 = 72$

20. $(x + 4)^2 = 768$

21. $(x - 10)^2 = 507$

22. $(x - 2)^2 = 112$

23. $(x + 8)^2 = 338$

24. $(x - 6)^2 + 320 = 0$

25. $(x + 2)^2 = 325$

26. $(x + 1)^2 = 675$

27. $(x - 4)^2 = -700$

28. $(x + 3)^2 - 252 = 0$

29. $5(x - 7)^2 = 144$

30. $6(x - 1)^2 = 324$

31. $3(x + 2)^2 = 49$

32. $7(x + 8)^2 = 625$

33. $12(x - 1)^2 = 361$

34. $18(x - 3)^2 = 121$

Completing the Square

Name _____

Date _____ Period _____

Complete each expression to create a perfect square trinomial.

1. $x^2 + 4x + \underline{\quad 4 \quad}$

2. $x^2 + 8x + \underline{\qquad}$

3. $x^2 - 20x + \underline{\qquad}$

4. $x^2 + 12x + \underline{\qquad}$

5. $x^2 + 26x + \underline{\qquad}$

6. $x^2 - 30x + \underline{\qquad}$

7. $x^2 - 18x + \underline{\qquad}$

8. $x^2 - 50x + \underline{\qquad}$

9. $x^2 + 9x + \underline{\qquad}$

10. $x^2 + 11x + \underline{\qquad}$

11. $x^2 + 31x + \underline{\qquad}$

12. $x^2 + 23x + \underline{\qquad}$

13. $x^2 - 21x + \underline{\qquad}$

14. $x^2 + 33x + \underline{\qquad}$

15. $x^2 + 19x + \underline{\qquad}$

16. $x^2 - 45x + \underline{\qquad}$

17. $x^2 + \frac{1}{3}x + \underline{\qquad}$

18. $x^2 - \frac{1}{4}x + \underline{\qquad}$

19. $x^2 + \frac{1}{8}x + \underline{\qquad}$

20. $x^2 - \frac{3}{10}x + \underline{\qquad}$

21. $x^2 + \frac{3}{4}x + \underline{\qquad}$

22. $x^2 - \frac{7}{8}x + \underline{\qquad}$

23. $x^2 - \frac{9}{10}x + \underline{\qquad}$

24. $x^2 + \frac{2}{7}x + \underline{\qquad}$

25. $x^2 + \frac{3}{5}x + \underline{\qquad}$

26. $x^2 - \frac{7}{10}x + \underline{\qquad}$

27. $x^2 - \frac{8}{7}x + \underline{\qquad}$

28. $x^2 - \frac{7}{4}x + \underline{\qquad}$

29. $x^2 + \frac{4}{9}x + \underline{\qquad}$

30. $x^2 + \frac{11}{5}x + \underline{\qquad}$

31. $x^2 - \frac{5}{7}x + \underline{\qquad}$

32. $x^2 - \frac{12}{7}x + \underline{\qquad}$

33. $x^2 - \frac{16}{7}x + \underline{\qquad}$

34. $x^2 - \frac{10}{7}x + \underline{\qquad}$

Developing Skills in Algebra   Book D

Completing the Square

Name _____

Date _____ Period _____

Complete each expression to create a perfect square trinomial.

1. $x^2 + 2x + \underline{1}$

2. $x^2 + 6x + \underline{}$

3. $x^2 - 14x + \underline{}$

4. $x^2 + 10x + \underline{}$

5. $x^2 + 24x + \underline{}$

6. $x^2 - 16x + \underline{}$

7. $x^2 - 22x + \underline{}$

8. $x^2 - 28x + \underline{}$

9. $x^2 + x + \underline{}$

10. $x^2 + 7x + \underline{}$

11. $x^2 + 13x + \underline{}$

12. $x^2 + 3x + \underline{}$

13. $x^2 - 25x + \underline{}$

14. $x^2 + 15x + \underline{}$

15. $x^2 + 17x + \underline{}$

16. $x^2 - 51x + \underline{}$

17. $x^2 + \frac{1}{2}x + \underline{}$

18. $x^2 - \frac{1}{5}x + \underline{}$

19. $x^2 + \frac{4}{5}x + \underline{}$

20. $x^2 - \frac{1}{7}x + \underline{}$

21. $x^2 + \frac{3}{7}x + \underline{}$

22. $x^2 - \frac{2}{3}x + \underline{}$

23. $x^2 - \frac{5}{4}x + \underline{}$

24. $x^2 + \frac{1}{10}x + \underline{}$

25. $x^2 + \frac{1}{9}x + \underline{}$

26. $x^2 - \frac{2}{5}x + \underline{}$

27. $x^2 - \frac{6}{7}x + \underline{}$

28. $x^2 - \frac{5}{8}x + \underline{}$

29. $x^2 + \frac{4}{7}x + \underline{}$

30. $x^2 + \frac{14}{9}x + \underline{}$

31. $x^2 - \frac{9}{4}x + \underline{}$

32. $x^2 - \frac{3}{8}x + \underline{}$

33. $x^2 - \frac{2}{9}x + \underline{}$

34. $x^2 - \frac{13}{5}x + \underline{}$

Quadratic Equations

Name _____

Date _____ Period _____

Solve by completing the square.

1. $x^2 + 2x = 12$ $x = -1 + \sqrt{13}$
  or $-1 - \sqrt{13}$

2. $x^2 + 2x = 21$

3. $x^2 + 2x = 7$

4. $x^2 + 2x = 17$

5. $x^2 - 4x = 13$

6. $x^2 - 4x = 25$

7. $x^2 + 4x = 20$

8. $x^2 + 4x = 23$

9. $x^2 + 6x = 12$

10. $x^2 + 6x = 14$

11. $x^2 - 6x = 23$

12. $x^2 - 6x = 31$

13. $x^2 - 8x = 21$

14. $x^2 - 8x = 13$

15. $x^2 + 8x = 11$

16. $x^2 + 8x = 8$

17. $x^2 + 12x + 7 = 0$

18. $x^2 + 12x + 12 = 0$

19. $x^2 - 20x = -14$

20. $x^2 - 14x = -13$

21. $x^2 + 3x = \dfrac{1}{4}$

22. $x^2 + 5x = \dfrac{3}{4}$

23. $x^2 + 7x + \dfrac{7}{4} = 0$

24. $x^2 + 13x + \dfrac{9}{4} = 0$

25. $x^2 + 11x = 13$

26. $x^2 + 9x = -12$

27. $x^2 - 5x - 10 = 0$

28. $x^2 - 7x + 15 = 0$

29. $x^2 + 3x = 3$

30. $x^2 + x = 7$

31. $x^2 + 5x - 7 = 0$

32. $x^2 + 6x - 5 = 0$

33. $x^2 - 3x + 10 = 0$

34. $x^2 - 4x + 3 = 0$

Quadratic Equations

Name _____

Date _____ Period _____

Solve by completing the square.

1. $x^2 + 2x = 5$ $x = -1 + \sqrt{6}$ or $-1 - \sqrt{6}$

2. $x^2 - 5x = 8$

3. $x^2 - 4x = 3$

4. $x^2 - 8x + 13 = 0$

5. $x^2 + 3x = 5$

6. $x^2 + 6x = 5$

7. $x^2 - 5x + 7 = 0$

8. $x^2 - 2x = 16$

9. $x^2 - 6x - 13 = 0$

10. $x^2 - 7x - 5 = 0$

11. $x^2 + 8x - 5 = 0$

12. $x^2 + 3x + 11 = 0$

13. $x^2 + 4x = 13$

14. $x^2 - 9x + 12 = 0$

15. $x^2 - 7x + 5 = 0$

16. $x^2 - 4x = 9$

17. $x^2 - 9x = \dfrac{5}{4}$

18. $x^2 + 10x + 12 = 0$

19. $x^2 + 2x = 11$

20. $x^2 + 5x = \dfrac{7}{4}$

21. $x^2 - 10x + 17 = 0$

22. $x^2 + 9x + 3 = 0$

23. $x^2 - 11x + \dfrac{1}{4} = 0$

24. $x^2 - 8x + 1 = 0$

25. $x^2 + 3x - 21 = 0$

26. $x^2 - 7x = \dfrac{3}{4}$

27. $x^2 + 5x - 16 = 0$

28. $x^2 - 2x = 17$

29. $x^2 + 12x + 9 = 0$

30. $x^2 - 6x - 13 = 0$

31. $x^2 - 4x + 2 = 0$

32. $x^2 - 5x - 7 = 0$

33. $x^2 + 7x + 4 = 0$

34. $x^2 + 3x - 2 = 0$

Quadratic Equations

Name _____

Date _____ Period _____

Solve by completing the square.

1. $2x^2 - 3x - 6 = 0$ $x = \dfrac{3 + \sqrt{57}}{4}$ or $\dfrac{3 - \sqrt{57}}{4}$

2. $2x^2 + 5x - 5 = 0$

3. $3x^2 + 12x - 2 = 0$

4. $2x^2 + 6x - 3 = 0$

5. $3x^2 + 9x + 5 = 0$

6. $5x^2 - 8x + 2 = 0$

7. $6x^2 - 10x + 3 = 0$

8. $4x^2 - 12x + 5 = 0$

9. $2x^2 - 9x + 6 = 0$

10. $3x^2 - 7x - 5 = 0$

11. $2x^2 + 10x - 1 = 0$

12. $5x^2 + 6x - 4 = 0$

13. $3x^2 + 11x + 6 = 0$

14. $4x^2 + 9x + 5 = 0$

15. $2x^2 + 7x - 6 = 0$

16. $6x^2 + 11x + 4 = 0$

17. $3x^2 + 10x + 6 = 0$

18. $6x^2 + 2x + 7 = 0$

19. $5x^2 + 9x - 6 = 0$

20. $7x^2 + 6x - 5 = 0$

21. $3x^2 - 10x + 1 = 0$

22. $2x^2 - 9x + 4 = 0$

23. $5x^2 - 6x + 7 = 0$

24. $7x^2 + 10x - 5 = 0$

25. $2x^2 + 11x + 7 = 0$

26. $4x^2 + 2x + 9 = 0$

27. $7x^2 - 9x + 2 = 0$

28. $3x^2 - 10x + 4 = 0$

29. $4x^2 + 7x + 7 = 0$

30. $2x^2 - 6x - 5 = 0$

31. $2x^2 - 13x - 5 = 0$

32. $5x^2 + 4x - 3 = 0$

33. $3x^2 + 2x - 9 = 0$

34. $6x^2 - 13x + 5 = 0$

Quadratic Equations

Name _____

Date _____ Period _____

Solve by completing the square.

1. $2x^2 + 3x = 7$ $x = \dfrac{-3 + \sqrt{65}}{4}$ or $\dfrac{-3 - \sqrt{65}}{4}$

2. $3x^2 + 5x = 1$

3. $3x^2 - 6x = -5$

4. $7x^2 - 4x = 3$

5. $5x^2 - 4x = 2$

6. $2x^2 + 7x = 7$

7. $7x^2 - 5x + 7 = 0$

8. $9x^2 - 11x = 9$

9. $2x^2 + 13x + 6 = 0$

10. $7x^2 + 6x + 1 = 0$

11. $9x^2 - 4x = 5$

12. $10x^2 + 4x - 5 = 0$

13. $2x^2 + 7x - 8 = 0$

14. $2x^2 + 5x + 9 = 0$

15. $3x^2 + 8x = 9$

16. $5x^2 - 6x + 1 = 0$

17. $7x^2 - 11x + 1 = 0$

18. $2x^2 - 3x - 3 = 0$

19. $2x^2 + 5x - 6 = 0$

20. $3x^2 + 7x - 9 = 0$

21. $9x^2 - 13x + 7 = 0$

22. $7x^2 - 8x + 1 = 0$

23. $5x^2 + 9x - 9 = 0$

24. $9x^2 + 11x - 7 = 0$

25. $10x^2 + 7x - 5 = 0$

26. $2x^2 - 2x + 5 = 0$

27. $2x^2 + 3x - 3 = 0$

28. $5x^2 + 13x + 6 = 0$

29. $3x^2 + 10x - 10 = 0$

30. $10x^2 - 11x + 1 = 0$

31. $2x^2 + 7x - 3 = 0$

32. $3x^2 + 5x + 3 = 0$

33. $2x^2 + 9x + 7 = 0$

34. $5x^2 - 10x - 7 = 0$

© 1984 by Dale Seymour Publications

54

Developing Skills in Algebra   Book D

Quadratic Equations

Name _____

Date _____ Period _____

Solve by the quadratic formula.

1. $x^2 + 4x - 8 = 0$ $\quad x = -2 + 2\sqrt{3}$ or $-2 - 2\sqrt{3}$

2. $x^2 - 6x + 4 = 0$

3. $x^2 - 16x + 19 = 0$

4. $x^2 - 6x + 6 = 0$

5. $x^2 + 10x + 17 = 0$

6. $x^2 - 6x - 3 = 0$

7. $x^2 + 5x + 13 = 0$

8. $x^2 - 8x - 47 = 0$

9. $x^2 - 7x + 12 = 0$

10. $x^2 + 6x - 41 = 0$

11. $x^2 + 2x - 17 = 0$

12. $x^2 - 7x + 17 = 0$

13. $x^2 + 12x - 4 = 0$

14. $x^2 + 16x - 35 = 0$

15. $x^2 + 10x - 3 = 0$

16. $x^2 - 8x + 12 = 0$

17. $x^2 - 3x + 9 = 0$

18. $x^2 + 2x - 1 = 0$

19. $x^2 + 6x - 243 = 0$

20. $x^2 + 4x + 1 = 0$

21. $x^2 + 6x + 5 = 0$

22. $x^2 + 2x + 3 = 0$

23. $x^2 + 6x - 81 = 0$

24. $x^2 + 8x - 2 = 0$

25. $x^2 + 16x - 122 = 0$

26. $x^2 + 8x + 11 = 0$

27. $x^2 - 16x - 96 = 0$

28. $x^2 - x - 6 = 0$

29. $x^2 - 10x + 1 = 0$

30. $x^2 - 6x - 11 = 0$

31. $x^2 + 6x - 39 = 0$

32. $x^2 + 10x - 7 = 0$

33. $x^2 - 10x + 22 = 0$

34. $x^2 - 20x - 8 = 0$

Developing Skills in Algebra   Book D

Quadratic Equations

Name _____

Date _____ Period _____

Solve by the quadratic formula.

1. $x^2 - 8x - 34 = 0$ $x =$ $4 + 5\sqrt{2}$ or $4 - 5\sqrt{2}$

2. $x^2 + 14x - 1 = 0$

3. $x^2 - 10x + 13 = 0$

4. $x^2 - 8x - 83 = 0$

5. $x^2 + 14x + 41 = 0$

6. $x^2 + 6x - 30 = 0$

7. $x^2 - 10x + 1 = 0$

8. $x^2 + 6x - 4 = 0$

9. $x^2 + 5x - 14 = 0$

10. $x^2 - 22x + 9 = 0$

11. $x^2 - 6x + 19 = 0$

12. $x^2 + 10x - 20 = 0$

13. $x^2 + 4x - 59 = 0$

14. $x^2 - 10x - 25 = 0$

15. $x^2 - 14x + 21 = 0$

16. $x^2 + 6x - 43 = 0$

17. $x^2 - 5x - 24 = 0$

18. $x^2 - 6x + 15 = 0$

19. $x^2 + 10x - 38 = 0$

20. $x^2 - 8x - 4 = 0$

21. $x^2 - 3x + 12 = 0$

22. $x^2 + 8x - 20 = 0$

23. $x^2 - 14x + 31 = 0$

24. $x^2 - 16x - 26 = 0$

25. $x^2 + 6x - 15 = 0$

26. $x^2 - 7x + 20 = 0$

27. $x^2 + 18x + 18 = 0$

28. $x^2 + 14x - 31 = 0$

29. $x^2 - 16x - 83 = 0$

30. $x^2 - 6x - 35 = 0$

31. $x^2 - 4x - 76 = 0$

32. $x^2 - 4x - 1 = 0$

33. $x^2 + 6x - 19 = 0$

34. $x^2 - 12x + 30 = 0$

56

Quadratic Equations

Name _____

Date _____ Period _____

Solve by the quadratic formula.

1. $4x^2 + 12x + 7 = 0$ $x = \dfrac{-3 + \sqrt{2}}{2}$ or $\dfrac{-3 - \sqrt{2}}{2}$

2. $25x^2 - 10x - 71 = 0$

3. $4x^2 - 16x + 11 = 0$

4. $9x^2 - 6x - 49 = 0$

5. $2x^2 + 10x + 11 = 0$

6. $4x^2 - 24x - 89 = 0$

7. $9x^2 - 6x - 79 = 0$

8. $20x^2 - 23x + 6 = 0$

9. $3x^2 + 7x - 8 = 0$

10. $7x^2 - 3x - 5 = 0$

11. $4x^2 - 28x + 37 = 0$

12. $9x^2 + 6x - 74 = 0$

13. $4x^2 + 16x - 109 = 0$

14. $3x^2 + 6x + 2 = 0$

15. $6x^2 - x + 1 = 0$

16. $5x^2 - 7x + 8 = 0$

17. $3x^2 + 2x + 5 = 0$

18. $9x^2 - 6x - 19 = 0$

19. $5x^2 - 3x - 3 = 0$

20. $18x^2 - 27x + 10 = 0$

21. $4x^2 + 8x - 23 = 0$

22. $9x^2 - 38x + 17 = 0$

23. $15x^2 - 8x + 1 = 0$

24. $6x^2 - 7x - 2 = 0$

25. $4x^2 + 4x - 5 = 0$

26. $2x^2 - 14x + 11 = 0$

27. $7x^2 - 3x + 2 = 0$

28. $4x^2 - 12x - 11 = 0$

29. $4x^2 - 28x + 31 = 0$

30. $4x^2 - 8x - 23 = 0$

31. $3x^2 - 7x - 9 = 0$

32. $5x^2 - 3x + 1 = 0$

33. $4x^2 + 4x - 17 = 0$

34. $4x^2 - 4x - 11 = 0$

Quadratic Equations

Name _____

Date _____ Period _____

Solve by the quadratic formula. $x =$

1. $4x^2 - 4x - 19 = 0$ $\dfrac{1 + 2\sqrt{5}}{2}$ or $\dfrac{1 - 2\sqrt{5}}{2}$

2. $2x^2 - 6x + 3 = 0$

3. $4x^2 + 20x - 25 = 0$

4. $9x^2 - 4x - 19 = 0$

5. $2x^2 + 7x - 13 = 0$

6. $10x^2 - x - 3 = 0$

7. $9x^2 - 36x - 2 = 0$

8. $5x^2 - 3x + 1 = 0$

9. $6x^2 + 5x - 6 = 0$

10. $2x^2 - 8x + 7 = 0$

11. $4x^2 - 20x + 23 = 0$

12. $9x^2 - 6x - 31 = 0$

13. $4x^2 - 4x - 7 = 0$

14. $4x^2 - 20x - 55 = 0$

15. $3x^2 - 2x + 7 = 0$

16. $3x^2 - 7x - 5 = 0$

17. $4x^2 + 28x + 41 = 0$

18. $9x^2 - 12x - 71 = 0$

19. $2x^2 - 7x - 11 = 0$

20. $4x^2 - 4x - 1 = 0$

21. $14x^2 - x - 4 = 0$

22. $7x^2 - 9x - 3 = 0$

23. $9x^2 - 18x - 41 = 0$

24. $4x^2 - 12x - 9 = 0$

25. $3x^2 - 5x - 10 = 0$

26. $7x^2 - 3x + 2 = 0$

27. $4x^2 - 16x + 13 = 0$

28. $5x^2 - 3x - 1 = 0$

29. $2x^2 - 2x - 1 = 0$

30. $7x^2 - 9x + 2 = 0$

31. $5x^2 - 8x + 11 = 0$

32. $9x^2 - 6x - 44 = 0$

33. $4x^2 + 4x - 47 = 0$

34. $2x^2 - 4x - 23 = 0$

The Discriminant

Name _____

Date _____ Period _____

Find $b^2 - 4ac$, then check the appropriate column for each equation.

	$b^2 - 4ac$	one double rat. root	two different rat. roots	two different irrat. roots	no real roots
1. $x^2 + x - 2 = 0$	9		✓		
2. $3x^2 - 22x + 7 = 0$					
3. $x^2 - 8x + 4 = 0$					
4. $2x^2 - 9x + 13 = 0$					
5. $4x^2 - 12x - 39 = 0$					
6. $4x^2 - 12x + 9 = 0$					
7. $4x^2 - 20x + 5 = 0$					
8. $15x^2 + 11x - 12 = 0$					
9. $5x^2 - 7x + 3 = 0$					
10. $6x^2 - 11x - 21 = 0$					
11. $4x^2 + 20x + 25 = 0$					
12. $9x^2 + 30x + 25 = 0$					
13. $2x^2 + 6x - 9 = 0$					
14. $4x^2 - 12x - 19 = 0$					
15. $6x^2 + 10x - 5 = 0$					
16. $16x^2 - 8x + 14 = 0$					

Developing Skills in Algebra   Book D

The Discriminant

Name _____

Date _____ Period _____

Find $b^2 - 4ac$, then check the appropriate column for each equation.

	$b^2 - 4ac$	one double rat. root	two different rat. roots	two different irrat. roots	no real roots
1. $12x^2 - x - 6 = 0$	289		✓		
2. $5x^2 - 31x + 6 = 0$					
3. $x^2 - 12x + 33 = 0$					
4. $3x^2 - 8x + 7 = 0$					
5. $x^2 - 8x + 16 = 0$					
6. $9x^2 - 30x + 23 = 0$					
7. $4x^2 + 20x - 23 = 0$					
8. $9x^2 + 42x + 49 = 0$					
9. $4x^2 + 28x + 37 = 0$					
10. $6x^2 - 7x - 5 = 0$					
11. $x^2 + 10x - 2 = 0$					
12. $2x^2 - 5x + 4 = 0$					
13. $9x^2 - 24x - 11 = 0$					
14. $3x^2 - 29x + 40 = 0$					
15. $25x^2 + 90x + 81 = 0$					
16. $6x^2 - 12x + 10 = 0$					

60

Quadratic Equations

Name _____

Date _____ Period _____

Solve.

$x =$

1. $x^2 - 6x - 31 = 0$ $3 + 2\sqrt{10}$ or $3 - 2\sqrt{10}$ **2.** $x^2 + 2x - 4 = 0$

3. $x^2 - 6x + 4 = 0$ **4.** $x^2 - 8x + 12 = 0$

5. $x^2 - 10x + 25 = 0$ **6.** $x^2 - 4x + 1 = 0$

7. $x^2 - 20x + 92 = 0$ **8.** $x^2 + 9x + 3 = 0$

9. $x^2 + 5x - 66 = 0$ **10.** $x^2 + 12x + 31 = 0$

11. $x^2 - 7x + 13 = 0$ **12.** $x^2 + 6x + 9 = 0$

13. $x^2 - 10x - 38 = 0$ **14.** $x^2 - 6x + 5 = 0$

15. $x^2 - 26x + 94 = 0$ **16.** $x^2 - 3x + 1 = 0$

17. $x^2 + 3x - 28 = 0$ **18.** $x^2 - 12x - 14 = 0$

19. $2x^2 + 2x - 37 = 0$ **20.** $9x^2 + 30x - 23 = 0$

21. $14x^2 - 17x + 4 = 0$ **22.** $7x^2 - 5x - 2 = 0$

23. $6x^2 - 5x + 2 = 0$ **24.** $9x^2 + 7x - 5 = 0$

25. $5x^2 - 3x - 2 = 0$ **26.** $4x^2 - 20x - 7 = 0$

27. $2x^2 - 6x - 33 = 0$ **28.** $9x^2 - 42x + 49 = 0$

29. $5x^2 - 4x - 7 = 0$ **30.** $3x^2 - 8x + 6 = 0$

31. $25x^2 + 30x + 9 = 0$ **32.** $12x^2 - 11x + 2 = 0$

33. $3x^2 + 8x - 13 = 0$ **34.** $7x^2 - 5x - 21 = 0$

     Developing Skills in Algebra   Book D

Quadratic Equations

Name _____

Date _____ Period _____

Solve.

1. $x^2 - x - 31 = 0$ $x =$ $\dfrac{1 + 5\sqrt{5}}{2}$ or $\dfrac{1 - 5\sqrt{5}}{2}$

2. $x^2 - 2x - 31 = 0$

3. $x^2 - 20x + 25 = 0$

4. $x^2 + 10x - 29 = 0$

5. $x^2 - 9x + 20 = 0$

6. $x^2 - 3x - 29 = 0$

7. $x^2 - 3x + 10 = 0$

8. $x^2 - 26x + 169 = 0$

9. $x^2 - 2x - 27 = 0$

10. $x^2 - x - 11 = 0$

11. $x^2 - 6x - 9 = 0$

12. $x^2 + 5x - 25 = 0$

13. $x^2 + 12x + 36 = 0$

14. $x^2 - 5x - 14 = 0$

15. $x^2 - 4x - 23 = 0$

16. $x^2 + 2x + 8 = 0$

17. $x^2 - 8x - 29 = 0$

18. $x^2 + 6x - 3 = 0$

19. $4x^2 - 24x - 89 = 0$

20. $4x^2 - 16x - 11 = 0$

21. $9x^2 - 30x + 13 = 0$

22. $6x^2 - 5x - 10 = 0$

23. $16x^2 - 24x + 9 = 0$

24. $3x^2 - 7x + 11 = 0$

25. $2x^2 + 2x - 1 = 0$

26. $9x^2 - 12x - 8 = 0$

27. $2x^2 - 17x + 33 = 0$

28. $5x^2 - 8x - 2 = 0$

29. $7x^2 - 9x + 5 = 0$

30. $4x^2 + 4x - 17 = 0$

31. $2x^2 - 10x - 25 = 0$

32. $9x^2 + 12x + 4 = 0$

33. $3x^2 - 6x - 13 = 0$

34. $5x^2 - 16x + 11 = 0$

Name _____

Date _____ Period _____

Solve.

1. The length of a rectangle is 5 cm more than the width. The area of the rectangle is 84 cm². Find the dimensions of the rectangle. $w(w+5) = 84$

   $w = width$

   $w + 5 = length$

   $w^2 + 5w = 84$

   $w = \dfrac{-5 \pm \sqrt{25 + 336}}{2} = \dfrac{-5 \pm 19}{2} = -12 \text{ or } 7$

   The length is 12 cm and the width is 7 cm.

2. The length of a rectangle is 3 cm more than twice the width. The area is 1890 cm². Find the length and the width of the rectangle.

3. The base of a triangle is 2 m less than the altitude and the area of the triangle is 364 m². Find the altitude of the triangle.

4. In an isosceles triangle, the base is 4 cm less than the length of one of the equal sides, and the altitude is 9 cm more than the base. The area of the triangle is 810 cm². Find the length of the base and the altitude of the triangle.

5. A framed picture has length 35 cm and width 25 cm. The picture itself has area 375 cm². How far is it from the edge of the picture to the edge of the frame if this distance is uniform around the picture?

6. A manufacturer makes rectangular table tops measuring 24 cm by 36 cm. He wishes to make a table top that has twice the area of the ones he now makes, by adding the same number of centimeters to both the length and the width. Find the dimensions of the table tops he should manufacture.

7. A box without a top is to be made by cutting squares measuring 8 cm on a side from each corner of a square piece of tin and folding up the sides. Find the dimensions of the box, if it is to have a volume of 9248 cm³.

8. If Darrell and Ginny work together, they can finish a job in 2 h 55 min. Working alone, Ginny can finish the job in two hours less time than it would take Darrell. Find the time it would take each of them working alone to do the job.

Developing Skills in Algebra   Book D

Word Problems

Name _____

Date _____ Period _____

Solve.

1. The length of a rectangle is 7 cm more than the width. The area of the rectangle is 120 cm². Find the dimensions of the rectangle. *w (w + 7) = 120*   *The width is 8cm*
w = width   *w² + 7w = 120*   *and the length*
w + 7 = length   *is 15 cm.*
$$w = \frac{-7 \pm \sqrt{49 + 480}}{2} = \frac{-7 \pm 23}{2} = 8 \text{ or } -15$$

2. The length of a rectangle is 5 m more than three times the width. The area is 232 m². Find the length and the width of the rectangle.

3. The base of a triangle is 4 cm less than the altitude and the area of the triangle is 198 cm². Find the altitude of the triangle.

4. In an isosceles triangle, the base is 2 cm less than the length of one of the equal sides, and the altitude is 7 cm more than the base. The area of the triangle is 400 cm². Find the length of the base and the altitude of the triangle.

5. A framed picture has length 50 cm and width 40 cm. The picture itself has area 1344 cm². How far is it from the edge of the picture to the edge of the frame if this distance is uniform around the picture?

6. A manufacturer makes rectangular table tops measuring 30 cm by 36 cm. He wishes to make a table top that has 3 times the area of the ones he now makes, by adding the same number of centimeters to both the length and the width. Find the dimensions of the table tops he should manufacture.

7. A box without a top is to be made by cutting squares measuring 5 cm on a side from each corner of a square piece of tin and folding up the sides. Find the dimensions of the box, if it is to have a volume of 7220 cm³.

8. If Bob and Vera work together, they can finish a job in fifty hours. Working alone, Vera can finish the job in four hours less time than it would take Bob. Find the time it would take each of them working alone to do the job.

Word Problems

Name _____

Date _____ Period _____

Solve.

1. The length of a rectangle is 5 cm less than the diagonal and the width is 7 cm less than the diagonal. If the area is 195 cm^2, how long is the diagonal?

$$d = \frac{12 \pm \sqrt{144 + 640}}{2}$$

d = diagonal $(d - 7)(d - 5) = 195$ $= \frac{12 \pm 28}{2}$ The diagonal

$d - 7$ = width $d^2 - 12d + 35 = 195$ is 20 cm long.

$d - 5$ = length $d^2 - 12d - 160 = 0$ $= 20$ or -8

2. The length of the diagonal of a rectangle is 3 cm more than the width and the width is 9 cm less than the length. If the area is 580 cm^2, find the length of the rectangle.

3. The area of an equilateral triangle is $16\sqrt{3}$ square units. Find the altitude.

4. In an isosceles triangle, the base is 5 m longer than the length of one of the equal sides, and the altitude is 2 m less than the base. The area of the triangle is 391.5 m^2. Find the length of the base and the altitude of the triangle.

5. A picture which has length 18 cm and width 24 cm is framed. The total area enclosed by the frame is 720 cm^2. Find the dimensions of the frame.

6. James dug a flower bed 5 m by 10 m. He wished to dig another flower bed with three times the area of the first one by increasing each dimension of the first one by the same number of meters. Find the dimensions of the second flower bed.

7. A pan without a top is to be made by cutting squares measuring 3 cm on a side from each corner of a square piece of aluminum and folding up the sides. Find the area of the original piece of aluminum if the volume of the pan is to be 432 cm^3.

8. If Ty and Michael work together, they can overhaul the engine in their boat in 4 h 33 min. If it takes Michael 1 h less than twice as long as it takes Ty to do the job, find the time it would take each of them working alone to overhaul the engine.

Word Problems

Name _____

Date _____ Period _____

Solve.

1. The length of a rectangle is 2 m less than the diagonal and the width is 7 m less than the diagonal. If the area is 84 m², how long is the diagonal?

$$d = \frac{9 \pm \sqrt{81 + 280}}{2}$$

$d = diagonal$ $(d-2)(d-7) = 84$ $= \frac{9 \pm 19}{2}$ The diagonal is 14 m long.

$d-2 = length$ $d^2 - 9d + 14 = 84$

$d-7 = width$ $d^2 - 9d - 70 = 0$ $= 14 \text{ or } -5$

2. The length of the diagonal of a rectangle is 11 cm more than the width and the width is 3 cm less than the length. If the area is 990 cm², find the length of the rectangle.

3. The area of an equilateral triangle is $36\sqrt{3}$ square units. Find the altitude.

4. In an isosceles triangle, the base is 7 m longer than the length of one of the equal sides, and the altitude is 4 m less than the base. The area of the triangle is 142.5 m². Find the length of the base and the altitude of the triangle.

5. A picture which has length 12 cm and width 14 cm is framed. The total area enclosed by the frame is 288 cm². Find the dimensions of the frame.

6. Kelly dug a garden 18 m by 24 m. She wished to dig another garden with $1\frac{2}{3}$ times the area of the first one by increasing each dimension of the first one by the same number of meters. Find the dimensions of the second garden.

7. A pan without a top is to be made by cutting squares measuring 6 cm on a side from each corner of a square piece of aluminum and folding up the sides. Find the area of the original piece of aluminum if the volume of the pan is to be 1944 cm³.

8. If Bob and Bill work together, they can do a job in 3 h 44 min. If it takes Bob one hour more to do the job than it takes Bill, find the time it would take each of them working alone to do the job.

Complex Numbers

Name _____

Date _____ Period _____

Simplify.

1. $\sqrt{-1}$ i

2. $\sqrt{-4}$

3. $\sqrt{-16}$

4. $\sqrt{-25}$

5. $\sqrt{-50}$

6. $\sqrt{-72}$

7. $\sqrt{-80}$

8. $\sqrt{-98}$

9. $\sqrt{-108}$

10. $\sqrt{-125}$

11. $3i(2 + 4i)$

12. $2i(3 + 7i)$

13. $5i(3 - i)$

14. $4i(2 - 3i)$

15. $(5 + i)(2 + 5i)$

16. $(2 + 3i)(4 + 2i)$

17. $(2 - i)(5 + 3i)$

18. $(3 - i)(4 + 5i)$

19. $(5 - 5i)(5 + 5i)$

20. $(3 - 3i)(3 + 3i)$

21. $\dfrac{5}{1 + i}$

22. $\dfrac{4}{1 - i}$

23. $\dfrac{6i}{2 - 3i}$

24. $\dfrac{5i}{3 + 4i}$

25. $\dfrac{1 + i}{1 - i}$

26. $\dfrac{2 + 3i}{2 - 3i}$

Solve.

27. $x^2 + 1 = 0$ $\pm i$

28. $x^2 + 4 = 0$

29. $x^2 + 3x + 4 = 0$

30. $x^2 + 5x + 8 = 0$

31. $3x^2 + 2x + 1 = 0$

32. $2x^2 + 3x + 5 = 0$

Developing Skills in Algebra   Book D

Complex Numbers

Name _____

Date _____ Period _____

Simplify.

1. $\sqrt{-9}$ $3i$

2. $\sqrt{-36}$

3. $\sqrt{-81}$

4. $\sqrt{-49}$

5. $\sqrt{-20}$

6. $\sqrt{-27}$

7. $\sqrt{-48}$

8. $\sqrt{-96}$

9. $\sqrt{-225}$

10. $\sqrt{-150}$

11. $4i(6 + 7i)$

12. $5i(11 + 6i)$

13. $10i(12 + 3i)$

14. $9i(15 - 2i)$

15. $(10 + 4i)(5 + 6i)$

16. $(8 + 12i)(9 + 7i)$

17. $(5 - 15i)(3 + 4i)$

18. $(6 - 10i)(2 + 8i)$

19. $(8 - 8i)(8 + 8i)$

20. $(12 - 12i)(12 + 12i)$

21. $\dfrac{6}{2 + i}$

22. $\dfrac{7}{5 - i}$

23. $\dfrac{12i}{6 + 2i}$

24. $\dfrac{10i}{5 - 3i}$

25. $\dfrac{3 + 2i}{3 - 2i}$

26. $\dfrac{5 - 7i}{5 + 7i}$

Solve.

27. $x^2 + 16 = 0$ $\pm 4i$

28. $x^2 + 25 = 0$

29. $x^2 + 2x + 7 = 0$

30. $x^2 + 5x + 11 = 0$

31. $5x^2 + 12x + 14 = 0$

32. $6x^2 + 8x + 5 = 0$

Complex Numbers

Name _____

Date _____ Period _____

Simplify.

1. $\sqrt{-64}$ $8i$

2. $\sqrt{-100}$

3. $\sqrt{-121}$

4. $\sqrt{-196}$

5. $\sqrt{-175}$

6. $\sqrt{-63}$

7. $\sqrt{-90}$

8. $\sqrt{-147}$

9. $\sqrt{-144}$

10. $\sqrt{-180}$

11. $15i(12 - 14i)$

12. $14i(11 - 12i)$

13. $20i(10 + 18i)$

14. $25i(5 + 15i)$

15. $(9 + 7i)(15 + 5i)$

16. $(8 + 11i)(20 + 15i)$

17. $(15 - 10i)(9 + 8i)$

18. $(16 - 4i)(8 + 7i)$

19. $(14 - 14i)(14 + 14i)$

20. $(16 - 16i)(16 + 16i)$

21. $\dfrac{20}{8 + 5i}$

22. $\dfrac{25}{7 + 3i}$

23. $\dfrac{14i}{6 - 11i}$

24. $\dfrac{16i}{5 - 13i}$

25. $\dfrac{6 + i}{6 - i}$

26. $\dfrac{8 + i}{8 - i}$

Solve.

27. $x^2 + 36 = 0$ $\pm 6i$

28. $x^2 + 49 = 0$

29. $x^2 + x + 11 = 0$

30. $x^2 + x + 9 = 0$

31. $7x^2 + 6x + 4 = 0$

32. $8x^2 + 5x + 5 = 0$

Developing Skills in Algebra   Book D

Complex Numbers

Name _____

Date _____ Period _____

Simplify.

1. $\sqrt{-169}$ $13i$

2. $\sqrt{-225}$

3. $\sqrt{-256}$

4. $\sqrt{-400}$

5. $\sqrt{-200}$

6. $\sqrt{-128}$

7. $\sqrt{-192}$

8. $\sqrt{-245}$

9. $\sqrt{-52}$

10. $\sqrt{-600}$

11. $21i(16 - 10i)$

12. $23i(19 + 20i)$

13. $26i(14 + 30i)$

14. $24i(20 - 15i)$

15. $(16 + 9i)(9 + 8i)$

16. $(15 + 8i)(12 + 10i)$

17. $(30 - 8i)(12 + 9i)$

18. $(40 - 5i)(15 + 12i)$

19. $(20 - 20i)(20 + 20i)$

20. $(25 - 25i)(25 + 25i)$

21. $\dfrac{18}{2 - 2i}$

22. $\dfrac{22}{11 - 11i}$

23. $\dfrac{16i}{10 - 4i}$

24. $\dfrac{14i}{6 + 8i}$

25. $\dfrac{4 + 4i}{4 - 4i}$

26. $\dfrac{9 - 9i}{9 + 9i}$

Solve.

27. $x^2 + 64 = 0$ $\pm 8i$

28. $x^2 + 81 = 0$

29. $x^2 - 5x + 8 = 0$

30. $x^2 + 6x + 12 = 0$

31. $4x^2 + 6x + 3 = 0$

32. $8x^2 + 2x + 1 = 0$

70

Developing Skills in Algebra   Book D

Radical Equations

Name _____

Date _____ Period _____

Solve. Solutions must be checked.

1. $\sqrt{-3x - 2} - x = 0$ ϕ

2. $\sqrt{-10x - 16} - x = 0$

3. $-\sqrt{15x + 100} - x = 0$

4. $-\sqrt{8x + 65} - x = 0$

5. $\sqrt{22 - 9x} - x = 0$

6. $\sqrt{26 - 11x} - x = 0$

7. $-\sqrt{5x - 4} = x$

8. $-\sqrt{14 - 13x} = x$

9. $-\sqrt{-7x - 6} = x$

10. $-\sqrt{-11x - 10} = x$

11. $\sqrt{33 - 8x} = x$

12. $\sqrt{18 - 7x} = x$

13. $\sqrt{2x + 18} + 3 = x$

14. $\sqrt{8x + 17} - 1 = x$

15. $-\sqrt{3x + 19} + 3 = x$

16. $-\sqrt{-13x - 94} - 4 = x$

17. $\sqrt{18x - 119} + 3 = x$

18. $\sqrt{33x - 29} - 7 = x$

19. $-\sqrt{-14x - 161} - x = 8$

20. $-\sqrt{-9x - 45} - x = 9$

21. $\sqrt{25} - 12 = x$

22. $\sqrt{9} - 10 = x$

23. $-\sqrt{13x - 129} - x = 5$

24. $-\sqrt{19x + 264} - x = 6$

25. $\sqrt{x + 2} - \sqrt{2x + 5} = 0$

26. $\sqrt{x - 1} - \sqrt{2x - 6} = 0$

27. $\sqrt{3x - 10} + \sqrt{x + 8} = 0$

28. $\sqrt{5x - 18} + \sqrt{3x + 10} = 0$

29. $\sqrt{4x + 13} - \sqrt{x - 14} = 0$

30. $\sqrt{7x + 2} - \sqrt{4x - 4} = 0$

31. $\sqrt{9x - 4} - \sqrt{4x + 6} = 0$

32. $\sqrt{8x + 11} - \sqrt{7x + 3} = 0$

33. $\sqrt{x + 8} - \sqrt{x - 3} = 2$

34. $\sqrt{x - 7} - \sqrt{x + 2} = 3$

Radical Equations

Name _____

Date _____ Period _____

Solve. Solutions must be checked.

1. $-\sqrt{-10x - 25} - x = 0$ $x = -5$

2. $-\sqrt{32 - 4x} - x = 0$

3. $\sqrt{75 - 10x} - x = 0$

4. $\sqrt{-9x - 18} - x = 0$

5. $-\sqrt{14x + 32} - x = 0$

6. $-\sqrt{12x + 108} - x = 0$

7. $\sqrt{-7x - 12} + x = 0$

8. $\sqrt{-20x - 36} + x = 0$

9. $-\sqrt{11x + 42} = x$

10. $-\sqrt{13x + 48} = x$

11. $\sqrt{15x - 56} = x$

12. $\sqrt{17x - 52} = x$

13. $\sqrt{13x - 14} - 2 = x$

14. $\sqrt{14x - 3} - 3 = x$

15. $-\sqrt{-7x + 23} - 1 = x$

16. $-\sqrt{-x + 85} - 5 = x$

17. $\sqrt{-x - 9} - 11 = x$

18. $\sqrt{-5x - 46} - 12 = x$

19. $-\sqrt{5x + 235} - x = -13$

20. $-\sqrt{50x + 319} - x = 16$

21. $\sqrt{16} + 8 = x$

22. $\sqrt{49} - 7 = x$

23. $\sqrt{-16x + 377} - x = -17$

24. $-\sqrt{62x + 504} - x = 20$

25. $\sqrt{10x + 1} - \sqrt{8x - 1} = 0$

26. $\sqrt{12x + 11} - \sqrt{8x + 19} = 0$

27. $\sqrt{3x - 1} - \sqrt{3x + 1} = 0$

28. $\sqrt{5x - 4} - \sqrt{5x + 6} = 0$

29. $\sqrt{13x + 13} - \sqrt{8x - 7} = 0$

30. $\sqrt{17x - 11} - \sqrt{4x - 11} = 0$

31. $\sqrt{21x + 3} - \sqrt{17x - 1} = 0$

32. $\sqrt{19x - 5} - \sqrt{21x - 7} = 0$

33. $\sqrt{x + 10} - \sqrt{x + 6} = 4$

34. $\sqrt{x + 9} - \sqrt{x - 2} = 1$

Higher Order Radicals

Name _____

Date _____ Period _____

Simplify. Variables represent non-negative numbers.

1. $\sqrt[3]{8}$ *2*

2. $\sqrt[3]{27}$

3. $\sqrt[3]{54}$

4. $\sqrt[3]{16}$

5. $\sqrt[3]{-343}$

6. $\sqrt[3]{-125}$

7. $\sqrt[4]{16}$

8. $\sqrt[4]{81}$

9. $\sqrt[4]{64}$

10. $\sqrt[4]{3125}$

11. $\sqrt[5]{32}$

12. $\sqrt[5]{3125}$

13. $\sqrt[5]{-243}$

14. $\sqrt[5]{-256}$

15. $\sqrt[3]{125x^{10}y^4}$

16. $\sqrt[3]{27a^7b^9}$

17. $\sqrt[4]{16p^5q^7}$

18. $\sqrt[4]{256r^{10}t^2}$

Solve.

19. $x^3 + 27 = 0$ *x = −3*

20. $x^3 + 8 = 0$

21. $x^3 + 32 = 0$

22. $x^3 + 1296 = 0$

23. $x^4 - 81 = 0$

24. $x^4 - 2401 = 0$

25. $x^4 - 16 = 0$

26. $x^4 - 1296 = 0$

27. $x^5 - 32 = 0$

28. $x^5 - 1024 = 0$

29. $x^4 + 3x^2 + 2 = 0$

30. $x^4 + 7x^2 + 12 = 0$

31. $x^4 - 5x^2 - 6 = 0$

32. $x^4 + 2x^2 - 48 = 0$

Developing Skills in Algebra   Book D

Name _____

Date _____ **Period** _____

Simplify. Variables represent non-negative numbers.

1. $\sqrt[3]{512}$ *8*

2. $\sqrt[3]{125}$

3. $\sqrt[3]{81}$

4. $\sqrt[3]{1296}$

5. $\sqrt[3]{-64}$

6. $\sqrt[3]{-343}$

7. $\sqrt[4]{625}$

8. $\sqrt[4]{2401}$

9. $\sqrt[4]{243}$

10. $\sqrt[4]{7776}$

11. $\sqrt[5]{1024}$

12. $\sqrt[5]{243}$

13. $\sqrt[5]{-729}$

14. $\sqrt[5]{-64}$

15. $\sqrt[3]{64x^7y^{10}}$

16. $\sqrt[3]{8a^{12}b^{11}}$

17. $\sqrt[4]{81m^{15}n^{10}}$

18. $\sqrt[4]{2401s^{12}t^7}$

Solve.

19. $x^3 + 64 = 0$ *x = − 4*

20. $x^3 + 343 = 0$

21. $x^3 + 729 = 0$

22. $x^3 + 7776 = 0$

23. $x^4 - 256 = 0$

24. $x^4 - 4096 = 0$

25. $x^4 - 625 = 0$

26. $x^4 - 6561 = 0$

27. $x^5 - 243 = 0$

28. $x^5 - 7776 = 0$

29. $x^4 + 6x^2 + 8 = 0$

30. $x^4 + 12x^2 + 27 = 0$

31. $x^4 - 3x^2 - 28 = 0$

32. $x^4 + 5x^2 - 14 = 0$

Introduction to Functions

Name _____

Date _____ Period _____

State whether or not each set is a function. Answer *yes* or *no*.

1. $\{(4, 3), (-2, 10), (5, -6), (10, 7)\}$ *yes*

2. $\{(-3, -6), (-5, 10), (-1, 2), (0, 0)\}$

3. $\{(2, 7), (3, 7), (5, 7), (6, 7)\}$

4. $\{(7, 2), (7, 3), (7, 4), (7, 5)\}$

5. $\{(-5, 3), (6, 5), (3, 2), (10, 3)\}$

6. $\{(-7, 4), (8, 12), (9, 12), (6, 13)\}$

7. $\{(8, 6), (9, -3), (12, 5), (6, -3)\}$

8. $\{(-8, 2), (3, -1), (-6, -3), (7, -1)\}$

9. $\{(11, 5), (2, 7), (-3, 8), (-3, 10)\}$

10. $\{(6, 4), (-5, 2), (6, 7), (-8, 8)\}$

11. $\{(8, 6), (-5, 2), (0, 6), (-5, 1)\}$

12. $\{(9, 4), (3, 2), (-6, 4), (8, 7)\}$

13. $\{(x, y): y = 2x + 1\}$

14. $\{(x, y): y = 3x - 4\}$

15. $\{(x, y): y = -x + 7\}$

16. $\{(x, y): y = \frac{1}{2}x - 4\}$

17. _____
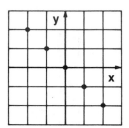

18. _____

19. _____
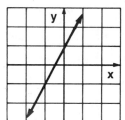

20. _____

21. _____
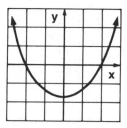

22. _____

Developing Skills in Algebra   Book D

Name _____

Date _____ Period _____

State whether or not each set is a function. Answer *yes* or *no*.

1. $\{(2, 1), (-5, 12), (8, -3), (11, 5)\}$ *yes*

2. $\{(-4, -2), (-4, 11), (-4, 6), (5, 5)\}$

3. $\{(3, 9), (4, 12), (5, 15), (6, 18)\}$

4. $\{(3, 5), (9, 5), (8, 5), (10, 5)\}$

5. $\{(-8, 4), (6, -3), (3, -\frac{3}{2}), (10, -5)\}$

6. $\{(5, -7), (3, -1), (8, 10), (5, 11)\}$

7. $\{(2, 7), (9, 28), (12, 37), (6, 19)\}$

8. $\{(-5, 3), (-5, 1), (-5, -2), (-5, 7)\}$

9. $\{(10, 0), (2, 0), (-6, 0), (15, 0)\}$

10. $\{(8, 3), (-7, 4), (5, 2), (-7, 4)\}$

11. $\{(1, 5), (-5, -19), (0, 1), (3, 13)\}$

12. $\{(7, 3), (1, 8), (-5, -2), (8, -1)\}$

13. $\{(x, y): y = -5x + 3\}$

14. $\{(x, y): y = \frac{1}{2}x + 8\}$

15. $\{(x, y): y = 2x - 1\}$

16. $\{(x, y): y = \frac{3}{4}x - 5\}$

17. _____
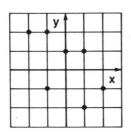

18. _____
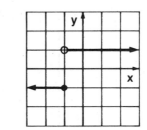

19. _____
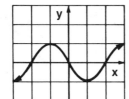

20. _____
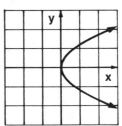

21. _____
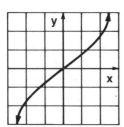

22. _____
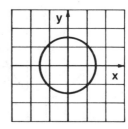

Given $\{(x, f(x)): f(x) = 2x + 3\}$, evaluate these expressions.

1. $f(3)$ *9*

2. $f(6)$

3. $f(-2)$

4. $f(-8)$

5. $f(10)$

6. $f(13)$

Given $\{(x, f(x)): f(x) = -6x + 5\}$, evaluate these expressions.

7. $f(4) + f(2)$

8. $f(5) + f(0)$

9. $f(6) - f(7)$

10. $f(11) - f(6)$

11. $f(-10) \cdot f(-5)$

12. $f(-6) \cdot f(-2)$

Given $\{(x, f(x)): f(x) = |x|\}$, evaluate these expressions.

13. $f(-7)$

14. $f(7)$

15. $f(15)$

16. $f(-15)$

17. $f(13)$

18. $f(-13)$

Given $\{(x, f(x)): f(x) = |x| + 6\}$, evaluate these expressions.

19. $f(-3)$

20. $f(3)$

21. $f(-10)$

22. $f(10)$

23. $f(7)$

24. $f(-7)$

Given $\{(x, f(x)): f(x) = x^2 + 4\}$, evaluate these expressions.

25. $f(-1)$

26. $f(1)$

27. $\dfrac{f(5) - f(3)}{5 - 3}$

28. $\dfrac{f(7) - f(2)}{7 - 2}$

Developing Skills in Algebra   Book D

Functional Notation

Name _____

Date _____ Period _____

Given $\{(x, f(x)): f(x) = -4x + 2\}$, evaluate these expressions.

1. $f(6)$ -22 **2.** $f(10)$

3. $f(2)$ **4.** $f(-13)$

5. $f(9)$ **6.** $f(-5)$

Given $\{(x, f(x)): f(x) = \frac{2}{3}x + 3\}$, evaluate these expressions.

7. $f(6) + f(12)$ **8.** $f(-3) + f(-12)$

9. $f(15) - f(3)$ **10.** $f(-6) - f(9)$

11. $f(21) \cdot f(0)$ **12.** $f(-24) \cdot f(24)$

Given $\{(x, f(x)): f(x) = 6 - |x|\}$, evaluate these expressions.

13. $f(10)$ **14.** $f(-10)$

15. $f(-5)$ **16.** $f(5)$

17. $f(-8)$ **18.** $f(8)$

Given $\{(x, f(x)): f(x) = |x| - 3\}$, evaluate these expressions.

19. $f(0)$ **20.** $f(5)$

21. $f(-2)$ **22.** $f(-4)$

23. $f(9)$ **24.** $f(2)$

Given $\{(x, f(x)): f(x) = x^2 + 3x - 1\}$, evaluate these expressions.

25. $f(2)$ **26.** $f(-4)$

27. $\dfrac{f(-2) - f(0)}{-2 - 0}$ **28.** $\dfrac{f(-1) - f(-3)}{-1 - (-3)}$

Name _____

Date _____ Period _____

Graph each function on the grid provided.

1. $\{(x, f(x)]) \; f(x) = -2x + 3\}$

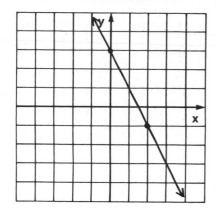

2. $\{(x, f(x)): f(x) = \frac{2}{3}x - 1\}$

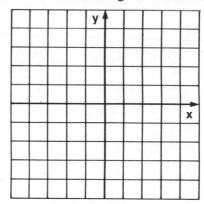

3. $\{(x, f(x)): f(x) = |x|\}$

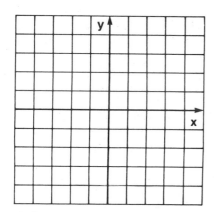

4. $\{(x, f(x)): f(x) = |x| + 2\}$

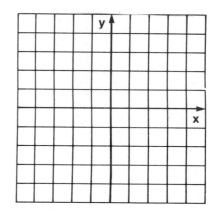

5. $\{(x, f(x)): f(x) = 2|x|\}$

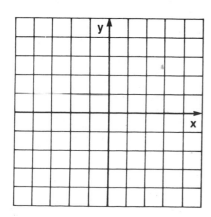

6. $\{(x, f(x)): f(x) = 3|x| - 1\}$

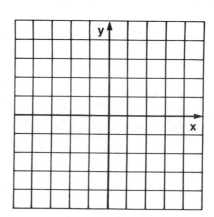

Developing Skills in Algebra   Book D

Name _____

Date _____ Period _____

Graph each function on the grid provided.

1. $\{(x, f(x)): f(x) = 4x - 2\}$

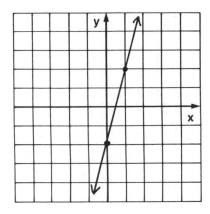

2. $\{(x, f(x)): f(x) = \dfrac{3}{4}x - 1\}$

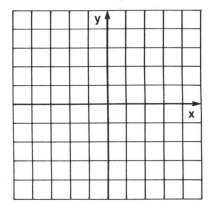

3. $\{(x, f(x)): f(x) = -|x|\}$

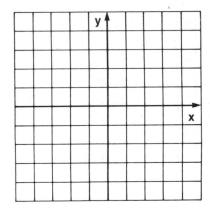

4. $\{(x, f(x)): f(x) = -|x| + 3\}$

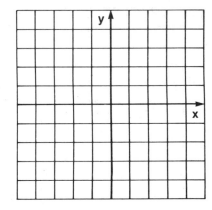

5. $\{(x, f(x)): f(x) = -3|x|\}$

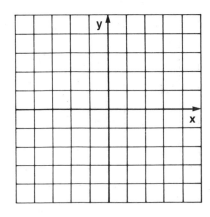

6. $\{(x, f(x)): f(x) = -2|x| - 1\}$

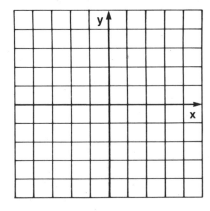

     Developing Skills in Algebra   Book D

Name _____

Date _____ Period _____

Complete the table, locate the points on the grid, and connect the points to make a smooth curve. Answer the questions about the graph.

Given $\{(x, y): y = x^2\}$

x	y
−4	16
−3	
−2	
−1	
0	
1	
2	
3	
4	

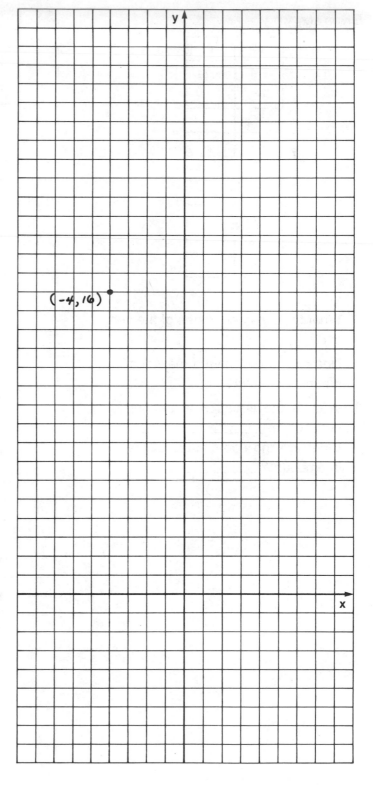

1. What are the coordinates of the vertex? _____

2. Does the curve have a maximum or a minimum point? _____

3. What is the equation of the line of symmetry? _____

4. What are the coordinates of the x-intercept(s)? _____

Graphing Quadratic Functions

Name _____

Date _____ Period _____

Complete the table, locate the points on the grid, and connect the points to make a smooth curve. Answer the questions about the graph.

Given $\{(x, y): y = -x^2\}$

x	y
−4	−16
−3	
−2	
−1	
0	
1	
2	
3	
4	

1. What are the coordinates of the vertex? _____

2. Does the curve have a maximum or a minimum point? _____

3. What is the equation of the line of symmetry? _____

4. What are the coordinates of the x-intercept(s)? _____

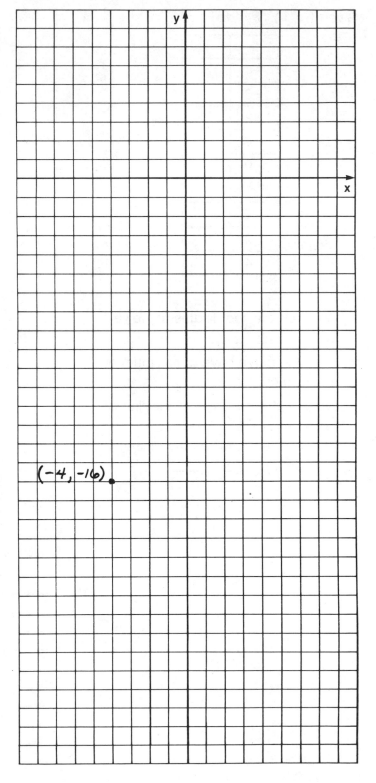

$(-4, -16)$

Developing Skills in Algebra   Book D

Graphing Quadratic Functions

Name _____

Date _____ Period _____

Complete the table, locate the points on the grid, and connect the points to make a smooth curve. Answer the questions about the graph.

Given $\{(x, y): y = 2x^2\}$

x	y
−4	32
−3	
−2	
−1	
0	
1	
2	
3	
4	

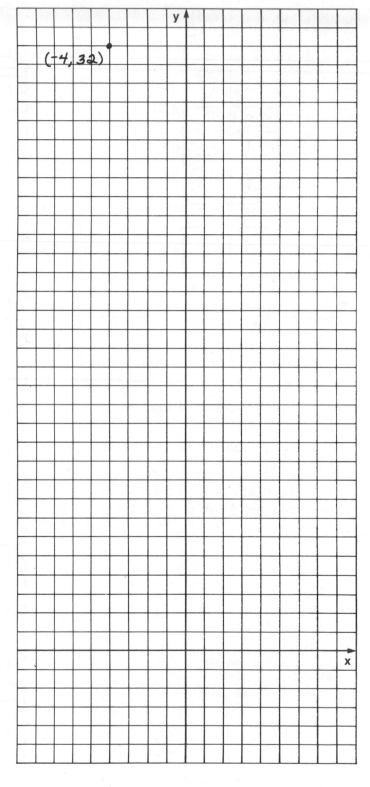

1. What are the coordinates of the vertex? _____

2. Does the curve have a maximum or a minimum point? _____

3. What is the equation of the line of symmetry? _____

4. What are the coordinates of the x-intercept(s)? _____

5. Does the graph open wider or narrower than the graph of $y = x^2$? _____

Developing Skills in Algebra   Book D

Graphing Quadratic Functions

Complete the table, locate the points on the grid, and connect the points to make a smooth curve. Answer the questions about the graph.

Given $\{(x, y): y = -3x^2\}$

x	y
−4	−48
−3	
−2	
−1	
0	
1	
2	
3	
4	

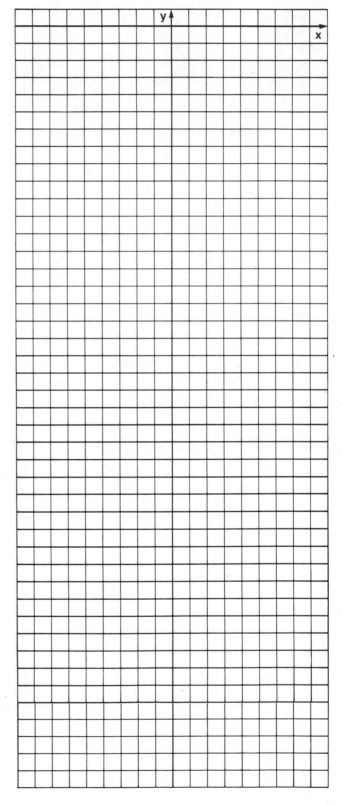

1. What are the coordinates of the vertex? _____

2. Does the curve have a maximum or a minimum point? _____

3. What is the equation of the line of symmetry? _____

4. What are the coordinates of the x-intercept(s)? _____

5. Does the graph open wider or narrower than the graph of $y = x^2$?

_____

Name _____

Date _____ Period _____

Graph each function by locating 5 points on each graph and joining them to form a smooth curve.

1. $\{(x, y): y = x^2 - 1\}$

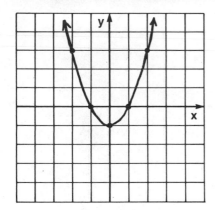

2. $\{(x, y): y = x^2 + 2\}$

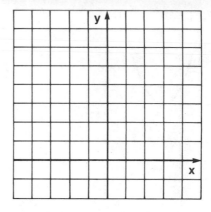

3. $\{(x, y): y = -x^2 + 3\}$

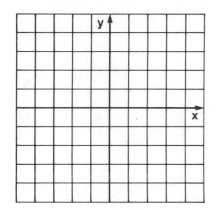

4. $\{(x, y): y = -x^2 - 1\}$

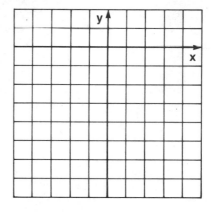

5. $\{(x, y): y = 2x^2 + 1\}$

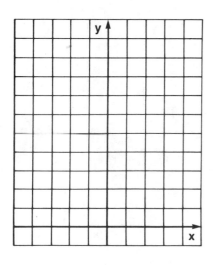

6. $\{(x, y): y = -\frac{1}{2}x^2 + 2\}$

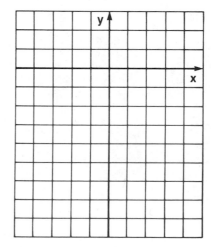

Developing Skills in Algebra   Book D

Name _____

Date _____ Period _____

Graph each function by locating 5 points on each graph and joining them to form a smooth curve.

1. $\{(x, y): y = (x - 1)^2\}$

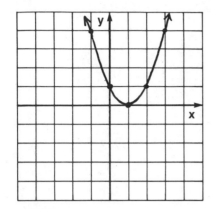

2. $\{(x, y): y = (x + 2)^2\}$

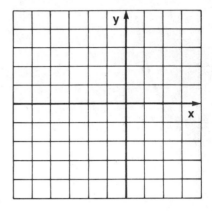

3. $\{(x, y): y = (x - 3)^2\}$

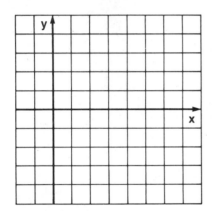

4. $\{(x, y): y = (x + 1)^2\}$

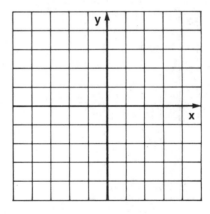

5. $\{(x, y): y = -2(x - 1)^2\}$

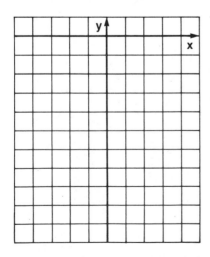

6. $\{(x, y): y = \frac{1}{2}(x + 2)^2\}$

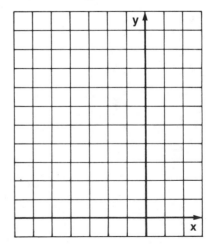

Developing Skills in Algebra   Book D

Analysis of Quadratic Functions

Name _____

Date _____ Period _____

List the correct information about the graphs of the functions specified by each equation.

	equation	vertex	maximum or minimum	line of symmetry	x-intercept	comparison to $x = y^2$
1.	$y = x^2 + 1$	$(0, 1)$	minimum	$x = 0$	none	same
2.	$y = 2x^2 + 3$					
3.	$y = \frac{1}{2}x^2 - 2$					
4.	$y = -6x^2 - 3$					
5.	$y = \frac{2}{3}x^2 + 5$					
6.	$y = (x + 3)^2$					
7.	$y = (x - 2)^2$					
8.	$y = -2(x + 1)^2$					
9.	$y = -\frac{2}{3}(x - 4)^2$					
10.	$y = 3(x - 5)^2$					

Developing Skills in Algebra   Book D

Analysis of Quadratic Functions

Name _____

Date _____ Period _____

List the correct information about the graphs of the functions specified by each equation.

	equation	vertex	maximum or minimum	line of symmetry	x-intercept	comparison to $x = y^2$
1.	$y = x^2 - 5$	$(0, -5)$	minimum	$x = 0$	$(\sqrt{5}, 0)$ $(-\sqrt{5}, 0)$	same
2.	$y = x^2 + 3$					
3.	$y = -2x^2 - 1$					
4.	$y = \frac{3}{4}x^2 + 5$					
5.	$y = -4x^2 + 2$					
6.	$y = (x - 2)^2$					
7.	$y = (x + 5)^2$					
8.	$y = 3(x - 4)^2$					
9.	$y = \frac{1}{3}(x + 2)^2$					
10.	$y = \frac{5}{3}(x - 1)^2$					

Graph the function specified by each equation by locating 5 points on each graph and joining them to form a smooth curve.

1. $y = (x - 1)^2 + 2$

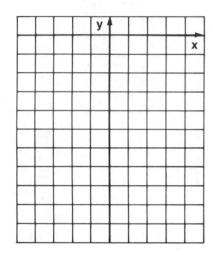

2. $y = (x + 1)^2 - 1$

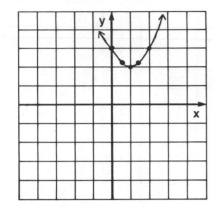

3. $y = -2(x + 1)^2 - 3$

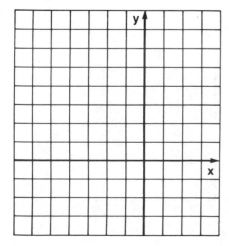

4. $y = -\frac{1}{2}(x - 2)^2 + 2$

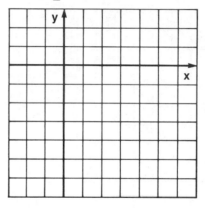

5. $y = \frac{1}{3}(x + 3)^2 - 4$

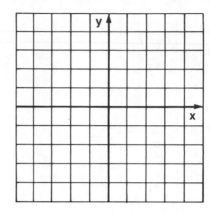

6. $y = 3(x - 1)^2 + 1$

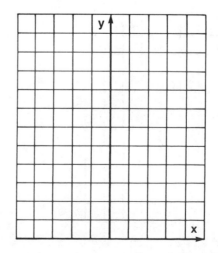

89

Name _____

Date _____ Period _____

Graph the function specified by each equation by locating 5 points on each graph and joining them to form a smooth curve.

1. $y = (x - 2)^2 + 1$

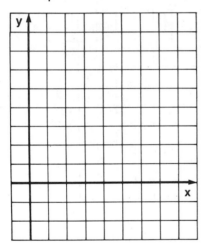

2. $y = -(x + 2)^2 - 1$

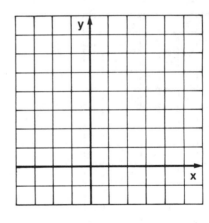

3. $y = \frac{3}{4}(x - 4)^2 - 3$

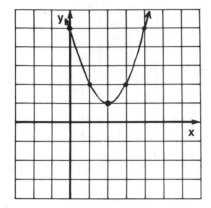

4. $y = 2(x - 1)^2 - 1$

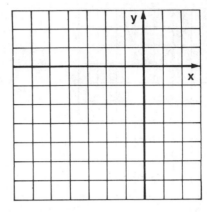

5. $y = -3(x + 2)^2 + 1$

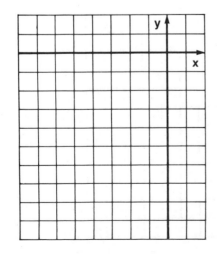

6. $y = -\frac{3}{2}(x - 2)^2 + 2$

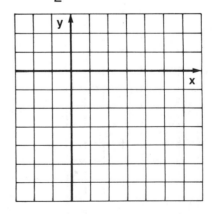

Developing Skills in Algebra   Book D

Analysis of Quadratic Functions

Name _____

Date _____ Period _____

List the correct information about the graphs of the functions specified by each equation.

	equation	vertex	maximum or minimum	line of symmetry	x-intercept	comparison to $x = y^2$
1.	$y = (x + 3)^2 - 4$	$(-3, -4)$	minimum	$x = -3$	$(-1, 0)$ $(-5, 0)$	same
2.	$y = (x - 1)^2 + 2$					
3.	$y = 2(x - 4)^2 + 3$					
4.	$y = -5(x + 1)^2 - 4$					
5.	$y = \frac{2}{3}(x - 2)^2 + 4$					
6.	$y = -\frac{5}{3}(x - 4)^2 + 1$					
7.	$y = 4(x - 1)^2 - 6$					
8.	$y = \frac{3}{8}(x + 1)^2 - 5$					
9.	$y = -\frac{2}{3}(x + 3)^2 - 4$					
10.	$y = -10(x - 1)^2 + 1$					

Developing Skills in Algebra   Book D

Analysis of Quadratic Functions

Name _____

Date _____ Period _____

List the correct information about the graphs of the functions specified by each equation.

	equation	vertex	maximum or minimum	line of symmetry	x-intercept	comparison to $x = y^2$
1.	$y = (x - 2)^2 + 3$	$(2, 3)$	minimum	$x = 2$	none	same
2.	$y = (x + 5)^2 - 1$					
3.	$y = -3(x + 6)^2 - 2$					
4.	$y = 6(x - 3)^2 + 1$					
5.	$y = -\frac{3}{4}(x - 5)^2 + 2$					
6.	$y = \frac{3}{2}(x - 5)^2 + 3$					
7.	$y = -\frac{4}{5}(x - 7)^2 + 2$					
8.	$y = -7(x + 3)^2 - 6$					
9.	$y = 5(x - 3)^2 + 7$					
10.	$y = \frac{7}{2}(x - 2)^2 + 9$					

Developing Skills in Algebra   Book D

Analysis of Quadratic Functions

Name _____

Date _____ Period _____

Find the symmetric point for each point given. Write an equation in the form $y = a(x - h)^2 + k$ which contains the vertex and the two symmetric points.

vertex	given point	symmetric point	equation
1. (0, 0)	(1, 1)	$(-1, 1)$	$y = x^2$ or $y = (x - 0)^2 + 0$
2. (0, 0)	(1, −1)	_____	_____
3. (0, 0)	(4, 2)	_____	_____
4. (0, 0)	(3, 10)	_____	_____
5. (0, −1)	(−3, 8)	_____	_____
6. (0, 5)	(4, 10)	_____	_____
7. (0, 2)	(1, 7)	_____	_____
8. (0, −3)	(−2, 5)	_____	_____
9. (5, 0)	(7, 7)	_____	_____
10. (−1, 0)	(2, 10)	_____	_____
11. (3, 0)	(5, 2)	_____	_____
12. (−2, 0)	(−1, 6)	_____	_____
13. (2, 3)	(4, 5)	_____	_____
14. (1, −2)	(3, 10)	_____	_____

Developing Skills in Algebra   Book D

Analysis of Quadratic Functions

Name _____

Date _____ Period _____

Find the symmetric point for each point given. Write an equation in the form $y = a(x - h)^2 + k$ which contains the vertex and the two symmetric points.

vertex	given point	symmetric point	equation
1. (0, 0)	(2, 3)	$(-2, 3)$	$y = \frac{3}{4}x^2$ or $y = \frac{3}{4}(x-0)^2 + 0$
2. (0, 0)	(3, 1)	_____	_____
3. (0, 0)	(1, −7)	_____	_____
4. (0, 0)	(3, −27)	_____	_____
5. (0, 4)	(2, 8)	_____	_____
6. (0, −2)	(3, −20)	_____	_____
7. (0, 1)	(5, 16)	_____	_____
8. (0, −6)	(2, −15)	_____	_____
9. (−4, 0)	(−2, 4)	_____	_____
10. (2, 0)	(1, 3)	_____	_____
11. (−5, 0)	(−2, 3)	_____	_____
12. (10, 0)	(5, −20)	_____	_____
13. (3, −5)	(6, −11)	_____	_____
14. (−1, 2)	(1, 30)	_____	_____

Developing Skills in Algebra   Book D

Find the zeros of the function specified by the given equation

1. $y = (x - 1)^2 - 5$ $x = 1 \pm \sqrt{5}$

2. $y = (x + 3)^2 - 2$

3. $y = (x + 2)^2 - 1$

4. $y = (x - 4)^2 - 3$

5. $y = (x - 6)^2 - 4$

6. $y = (x + 5)^2 - 2$

7. $y = (x - 3)^2 + 2$

8. $y = (x + 1)^2 + 3$

9. $y = 2(x - 4)^2 - 3$

10. $y = \frac{1}{2}(x + 3)^2 - 4$

11. $y = -\frac{2}{3}(x - 1)^2 + 6$

12. $y = -5(x - 2)^2 + 1$

13. $y = -3(x + 2)^2 + 7$

14. $y = -\frac{1}{5}(x - 1)^2 - 5$

15. $y = \frac{1}{4}(x - 1)^2 + 2$

16. $y = \frac{3}{4}(x - 1)^2 - 5$

17. $y = 6(x - 2)^2 - 5$

18. $y = -2(x + 1)^2 + 3$

19. $y = -3(x - 4)^2 - 2$

20. $y = \frac{1}{3}(x - 2)^2 + 5$

21. $y = \frac{2}{5}(x - 1)^2 - 1$

22. $y = 3(x - 2)^2 - 3$

23. $y = 3(x + 7)^2 + 4$

24. $y = -\frac{1}{2}(x - 2)^2 - 5$

25. $y = 4(x - 1)^2 - 3$

26. $y = 2(x - 4)^2 - 1$

27. $y = 5(x + 2)^2 - 6$

28. $y = -7(x - 4)^2 + 3$

29. $y = \frac{1}{2}(x - 3)^2 - 7$

30. $y = 4(x + 2)^2 - 5$

31. $y = -2(x - 1)^2 - 1$

32. $y = 3(x + 1)^2 - 6$

33. $y = \frac{3}{5}(x + 1)^2 - 3$

34. $y = -3(x + 2)^2 - 4$

Find the zeros of the function specified by the given equation.

1. $y = (x - 2)^2 - 3$ $x = 2 \pm \sqrt{3}$

2. $y = (x + 4)^2 - 1$

3. $y = (x + 6)^2 - 5$

4. $y = (x - 5)^2 - 6$

5. $y = (x - 1)^2 - 3$

6. $y = (x - 4)^2 - 2$

7. $y = (x + 3)^2 - 4$

8. $y = (x + 6)^2 - 3$

9. $y = 3(x - 3)^2 - 1$

10. $y = -\dfrac{4}{5}(x + 3)^2 + 2$

11. $y = 3(x - 3)^2 + 4$

12. $y = -3(x + 2)^2 - 3$

13. $y = -6(x + 1)^2 + 3$

14. $y = -2(x + 1)^2 + 4$

15. $y = -\dfrac{1}{2}(x - 2)^2 + 4$

16. $y = 3(x + 4)^2 + 5$

17. $y = \dfrac{2}{3}(x + 4)^2 - 5$

18. $y = \dfrac{5}{3}(x - 1)^2 - 2$

19. $y = -10(x - 1)^2 - 3$

20. $y = \dfrac{1}{2}(x - 2)^2 + 1$

21. $y = 7(x - 3)^2 - 2$

22. $y = -3(x - 1)^2 + 4$

23. $y = \dfrac{2}{3}(x + 2)^2 + 5$

24. $y = 2(x + 2)^2 - 7$

25. $y = 6(x - 1)^2 - 2$

26. $y = \dfrac{3}{4}(x - 3)^2 - 1$

27. $y = -5(x + 3)^2 + 7$

28. $y = -\dfrac{2}{3}(x - 4)^2 - 1$

29. $y = -\dfrac{5}{8}(x - 3)^2 - 4$

30. $y = 5(x - 1)^2 - 6$

31. $y = 10(x - 2)^2 - 3$

32. $y = \dfrac{3}{5}(x + 2)^2 - 1$

33. $y = 2(x - 1)^2 - 5$

34. $y = -3(x + 2)^2 + 4$

Completing the Square

Find the vertex of the graph of the function specified in the given equation by completing the square to write the equation in the form $y = a(x - h)^2 + k$.

1. $y = x^2 - 10x + 27$ *(5, 2)*

2. $y = x^2 + 6x + 4$

3. $y = x^2 + 2x - 6$

4. $y = x^2 - 8x + 19$

5. $y = x^2 - 4x - 2$

6. $y = x^2 + 20x + 95$

7. $y = x^2 + 14x + 40$

8. $y = x^2 - 6x + 11$

9. $y = x^2 - 10x + 21$

10. $y = x^2 + 16x + 61$

11. $y = x^2 - 6x + 15$

12. $y = x^2 - 4x - 1$

13. $y = 4x^2 - 8x + 5$

14. $y = -5x^2 + 50x - 119$

15. $y = -3x^2 - 12x - 13$

16. $y = \frac{1}{2}x^2 - 2x + 5$

17. $y = \frac{1}{3}x^2 - 2x + 7$

18. $y = 2x^2 + 4x$

19. $y = 3x^2 - 18x + 29$

20. $y = 6x^2 + 84x + 291$

21. $y = -2x^2 - 24x - 75$

22. $y = 7x^2 - 56x + 116$

23. $y = 8x^2 - 160x + 795$

24. $y = \frac{2}{5}x^2 - 4x + 4$

25. $y = \frac{3}{4}x^2 + 6x + 17$

26. $y = \frac{2}{3}x^2 - 4x + 13$

27. $y = -6x^2 + 72x - 223$

28. $y = 4x^2 - 64x + 263$

29. $y = 6x^2 + 36x + 62$

30. $y = \frac{2}{5}x^2 + 4x + 9$

31. $y = \frac{5}{3}x^2 + 30x + 138$

32. $y = 7x^2 - 98x + 352$

33. $y = \frac{4}{5}x^2 - 8x + 14$

34. $y = 5x^2 - 40x + 82$

Completing the Square

Name _____

Date _____ Period _____

Find the vertex of the graph of the function specified in the given equation by completing the square to write the equation in the form $y = a(x - h)^2 + k$.

1. $y = x^2 + 2x + 5$ $(-1, 4)$

2. $y = x^2 - 4x + 3$

3. $y = x^2 - 14x + 47$

4. $y = x^2 + 10x + 28$

5. $y = x^2 - 2x + 2$

6. $y = x^2 - 6x + 7$

7. $y = x^2 + 8x + 21$

8. $y = x^2 - 10x + 19$

9. $y = x^2 + 12x + 42$

10. $y = x^2 + 4x$

11. $y = x^2 - 8x + 13$

12. $y = x^2 - 12x + 39$

13. $y = 5x^2 + 20x + 17$

14. $y = -\frac{1}{2}x^2 + 4x - 10$

15. $y = -4x^2 + 8x - 5$

16. $y = 3x^2 + 6x + 6$

17. $y = \frac{1}{2}x^2 - 2x - 2$

18. $y = -5x^2 - 40x - 88$

19. $y = 2x^2 + 20x + 52$

20. $y = \frac{3}{4}x^2 - 3x + 10$

21. $y = \frac{2}{3}x^2 + 8x + 14$

22. $y = 4x^2 - 24x + 34$

23. $y = -6x^2 + 36x - 61$

24. $y = -\frac{1}{3}x^2 - 2x - 8$

25. $y = -7x^2 - 56x - 108$

26. $y = 6x^2 + 72x + 221$

27. $y = 5x^2 + 100x + 494$

28. $y = \frac{3}{5}x^2 - 6x + 20$

29. $y = -4x^2 - 24x - 43$

30. $y = -2x^2 + 4x + 6$

31. $y = 2x^2 - 44x + 246$

32. $y = \frac{3}{7}x^2 + 6x + 28$

33. $y = -3x^2 + 42x - 141$

34. $y = \frac{7}{12}x^2 - 7x + 24$

Name _____

Date _____ Period _____

Solve.

1. The height in feet of a rocket after t seconds is given by $h(t) = 160t - 16t^2$. Find the maximum height the rocket attains. $h(t) = -16t^2 + 224t$ The rocket attains
   $= -16(t^2 - 14t)$ a maximum height
   $= -16(t^2 - 14t + 49) + 784$ of 784 ft after 7 s.
   $= -16(t - 7)^2 + 784$

2. The height in feet of a rocket after t seconds is given by $h(t) = 256t - 16t^2$. After how many seconds does the rocket reach its maximum height?

3. Find the maximum rectangular area that can be enclosed by 452 m of fencing.

4. From all pairs of numbers whose sum is 256, find the pair whose product is greatest.

5. From all pairs of numbers whose sum is 20, find the pair for which the sum of their squares is least.

6. A piece of wire 24 cm long is cut into two pieces and each piece is bent to form a square. Find the length of each piece of wire in order to maximize the sum of the areas of the two squares.

7. In a 220 volt circuit having a resistance of 11 ohms, the power in watts, W, when a current I is flowing is given by the formula $W = 220I - 11I^2$. Find the maximum power that can be delivered in this circuit.

8. A building contractor estimates that his profit in dollars from building a highrise h stories high is given by $p = -3h^2 + 180h$. What building height will give him the most profit?

9. Suppose an open trough with a rectangular cross section is made from a sheet of metal by turning up x cm on each side. The function $A(x) = 20x - 2x^2$ determines the cross-sectional area of the trough. The volume of the trough is maximized when the cross-sectional area is at its maximum. For what value of x will the volume be maximized?

10. A printer will print a minimum order of 10 boxes of Valentine cards for $8 per box. For each order greater than 10, she will reduce the cost of the entire lot by $0.40 per box. The gross income to the printer for n boxes is given by the function $I(n) = 12n - 0.4n^2$. An order for what number of boxes would give the printer the maximum gross income?

Maximum/Minimum Word Problems

Name _____

Date _____ Period _____

Solve.

1. The height in feet of a rocket after t seconds is given by $h(t) = 224t - 16t^2$. Find the maximum height the rocket attains. $h(t) = -16t^2 - 160t$

The rocket attains a
$= -16(t^2 - 10t)$
maximum height of
$= -16(t^2 - 10t + 25) + 400$
400 ft after 5s.
$= -16(t-5)^2 + 400$

2. The height in feet of a rocket after t seconds is given by $h(t) = 192t - 16t^2$. After how many seconds does the rocket reach its maximum height?

3. Find the maximum rectangular area that can be enclosed by 600 meters of fencing.

4. From all pairs of numbers whose sum is 300, find the pair whose product is greatest.

5. From all pairs of numbers whose sum is 30, find the pair for which the sum of their squares is least.

6. A piece of wire 36 cm long is cut into two pieces and each piece is bent to form a square. Find the length of each piece of wire in order to maximize the sum of the areas of the two squares.

7. In a 110 volt circuit having a resistance of 5 ohms, the power in watts, W, when a current I is flowing is given by the formula $W = 110I - 5I^2$. Find the maximum power that can be delivered in this circuit.

8. A building contractor estimates that her profit in dollars from building a highrise h stories high is given by $p = -4h^2 + 160h$. What building height will give her the most profit?

9. Suppose an open trough with a rectangular cross section is made from a sheet of metal by turning up x cm on each side. The function $A(x) = 44x - 2x^2$ determines the cross-sectional area of the trough. The volume of the trough will be maximized when the cross-sectional area is at its maximum. For what value of x will the volume be maximized?

10. A printer will print a minimum order of 10 boxes of personalized stationery for $10 per box. For each order greater than 10, he will reduce the cost of the entire lot by $0.20 per box. The gross income to the printer for n boxes is given by the function $I(n) = 12n - 0.2n^2$. An order for what number of boxes would give the printer the maximum gross income?

ANSWERS

Page 1   Real Numbers

Answers may vary.

Page 2   Real Numbers

1. natural, whole, integer, rational   **2.** integer, rational
3. rational   **4.** rational   **5.** rational   **6.** irrational
7. natural, whole, integer, rational   **8.** natural, whole, integer, rational   **9.** rational   **10.** rational   **11.** irrational
12. rational   **13.** rational   **14.** natural, whole, integer, rational   **15.** whole, integer, rational   **16.** irrational
17. natural, whole, integer, rational   **18.** irrational
19. rational   **20.** rational   **21.** natural, whole, integer, rational   **22.** rational   **23.** rational   **24.** natural, whole, integer, rational

Page 3   Squares and Square Roots

1. 25   **2.** 1600   **3.** 289   **4.** 81   **5.** 841   **6.** 169
7. 43   **8.** 21   **9.** 35   **10.** 49   **11.** 25   **12.** 50   **13.** 33
14. 47   **15.** 11   **16.** 28   **17.** 36   **18.** 31   **19.** 66
20. 323   **21.** 328   **22.** 1125   **23.** 2/5   **24.** 6/7
25. 1/2   **26.** 3/4   **27.** 1.1   **28.** 4.2   **29.** 3.72
30. 17.8   **31.** $7ab$ **32.** $13c^4d^2$ **33.** $16a^n$ **34.** $21x^ry^r$

Page 4   Squares and Square Roots

1. 100   **2.** 361   **3.** 36   **4.** 1089   **5.** 1764   **6.** 196
7. 51   **8.** 24   **9.** 44   **10.** 28   **11.** 38   **12.** 48   **13.** 21
14. 15   **15.** 39   **16.** 35   **17.** 52   **18.** 85   **19.** 800
20. 1035   **21.** 384   **22.** 3180   **23.** 1/8   **24.** 4/15
25. 3/5   **26.** 5/7   **27.** 3.5   **28.** 5.6   **29.** 6.31
30. 11.3   **31.** $9b^2c$ **32.** $11r^2s^2t$ **33.** $25a^x$ **34.** $30a^nb^{2n}$

Page 5   Squares and Square Roots

1. 144   **2.** 49   **3.** 324   **4.** 225   **5.** 2704   **6.** 1296
7. 31   **8.** 41   **9.** 23   **10.** 47   **11.** 34   **12.** 56   **13.** 29
14. 24   **15.** 37   **16.** 25   **17.** 30   **18.** 98   **19.** 1978
20. 840   **21.** 162   **22.** 1862   **23.** 3/7   **24.** 8/17
25. 4/11   **26.** 5/9   **27.** 2.6   **28.** 7.5   **29.** 8.23
30. 16.2   **31.** $15m^4n^2$ **32.** $22x^7y^3$ **33.** $17m^n$
34. $26r^xs^{3x}$

Page 6   Squares and Square Roots

1. 400   **2.** 900   **3.** 64   **4.** 121   **5.** 2116   **6.** 1369
7. 39   **8.** 16   **9.** 22   **10.** 45   **11.** 27   **12.** 32   **13.** 9
14. 44   **15.** 34   **16.** 40   **17.** 51   **18.** 79   **19.** 170
20. 234   **21.** 3888   **22.** 240   **23.** 7/8   **24.** 3/7
25. 7/10   **26.** 8/15   **27.** 8.3   **28.** 6.5   **29.** 9.52
30. 21.4   **31.** $12x^3y^3$ **32.** $15a^4b^7$ **33.** $4v^a$
34. $32c^{3r}d^{5r}$

Page 7   Radical Expressions

1. $9\sqrt{2}$ **2.** $2\sqrt{2}$ **3.** $8\sqrt{3}$ **4.** $7\sqrt{5}$ **5.** $21\sqrt{7}$ **6.** $26\sqrt{3}$
7. $7\sqrt{2}$ **8.** $12\sqrt{5}$ **9.** $28\sqrt{6}$ **10.** $20\sqrt{11}$ **11.** $17\sqrt{3}$
12. $21\sqrt{6}$ **13.** $10\sqrt{7}$ **14.** $31\sqrt{2}$ **15.** $28\sqrt{3}$ **16.** $8\sqrt{5}$
17. $16\sqrt{13}$ **18.** $22\sqrt{6}$ **19.** $6\sqrt{3}$ **20.** $14\sqrt{10}$
21. $28\sqrt{11}$ **22.** $13\sqrt{6}$ **23.** $ab^2\sqrt{3}$ **24.** $m^2n\sqrt{2}$
25. $r^3s^4\sqrt{5}$ **26.** $ab^5\sqrt{7}$ **27.** $7a^4b\sqrt{10}$ **28.** $3a^2b^3\sqrt{2}$
29. $13a\sqrt{3a}$ **30.** $11r\sqrt{10rs}$ **31.** $12cd\sqrt{3c}$
32. $25r^3s^2\sqrt{5r}$ **33.** $14b^2c^3\sqrt{2bc}$ **34.** $7ab^2\sqrt{3ab}$

Page 8   Radical Expressions

1. $5\sqrt{2}$ **2.** $11\sqrt{3}$ **3.** $4\sqrt{2}$ **4.** $16\sqrt{2}$ **5.** $11\sqrt{6}$
6. $21\sqrt{5}$ **7.** $20\sqrt{13}$ **8.** $8\sqrt{2}$ **9.** $25\sqrt{10}$ **10.** $29\sqrt{5}$
11. $14\sqrt{6}$ **12.** $9\sqrt{15}$ **13.** $21\sqrt{17}$ **14.** $20\sqrt{3}$
15. $15\sqrt{5}$ **16.** $5\sqrt{10}$ **17.** $22\sqrt{22}$ **18.** $23\sqrt{7}$
19. $32\sqrt{3}$ **20.** $7\sqrt{6}$ **21.** $15\sqrt{30}$ **22.** $16\sqrt{11}$
23. $c^2d\sqrt{2}$ **24.** $r^2s^2\sqrt{5}$ **25.** $m^3n^2\sqrt{3}$ **26.** $r^3s^2\sqrt{10}$
27. $15m^2n^3\sqrt{13}$ **28.** $10x^3y\sqrt{5}$ **29.** $12c\sqrt{10c}$
30. $7m\sqrt{15mn}$ **31.** $14a^3b^2\sqrt{2ab}$ **32.** $13c^2d\sqrt{5d}$
33. $9rs^4\sqrt{5s}$ **34.** $17m^2n^2\sqrt{2mn}$

Page 9   Radical Expressions

1. $3\sqrt{3}$ **2.** $5\sqrt{3}$ **3.** $9\sqrt{7}$ **4.** $6\sqrt{5}$ **5.** $24\sqrt{6}$
6. $18\sqrt{15}$ **7.** $25\sqrt{3}$ **8.** $13\sqrt{2}$ **9.** $27\sqrt{2}$ **10.** $19\sqrt{10}$
11. $12\sqrt{17}$ **12.** $19\sqrt{2}$ **13.** $27\sqrt{15}$ **14.** $30\sqrt{19}$
15. $17\sqrt{5}$ **16.** $4\sqrt{13}$ **17.** $13\sqrt{22}$ **18.** $33\sqrt{3}$
19. $2\sqrt{30}$ **20.** $10\sqrt{15}$ **21.** $16\sqrt{3}$ **22.** $23\sqrt{2}$
23. $mn^2\sqrt{5}$ **24.** $p^2q\sqrt{3}$ **25.** $a^4b^3\sqrt{2}$ **26.** $x^5y\sqrt{11}$
27. $26a^3b\sqrt{2}$ **28.** $6r^2s^3\sqrt{7}$ **29.** $5x\sqrt{7x}$ **30.** $13c\sqrt{7c}$
31. $23x^3y^3\sqrt{5xy}$ **32.** $16pq^4\sqrt{5p}$ **33.** $12a^5b^2\sqrt{2ab}$
34. $18c^2d^5\sqrt{5cd}$

Page 10   Radical Expressions

1. $10\sqrt{5}$ **2.** $9\sqrt{6}$ **3.** $19\sqrt{3}$ **4.** $26\sqrt{2}$ **5.** $8\sqrt{7}$
6. $23\sqrt{6}$ **7.** $6\sqrt{10}$ **8.** $15\sqrt{15}$ **9.** $18\sqrt{3}$ **10.** $27\sqrt{5}$
11. $18\sqrt{7}$ **12.** $24\sqrt{2}$ **13.** $17\sqrt{10}$ **14.** $26\sqrt{15}$
15. $30\sqrt{17}$ **16.** $3\sqrt{30}$ **17.** $23\sqrt{11}$ **18.** $14\sqrt{5}$
19. $24\sqrt{7}$ **20.** $25\sqrt{6}$ **21.** $13\sqrt{15}$ **22.** $15\sqrt{22}$
23. $r^2s^2\sqrt{3}$ **24.** $v^2w\sqrt{5}$ **25.** $c^3d\sqrt{7}$ **26.** $a^2b^3\sqrt{15}$
27. $9r^2s\sqrt{3}$ **28.** $8m^3n\sqrt{2}$ **29.** $13y\sqrt{11y}$ **30.** $7a\sqrt{22ab}$
31. $30mn^4\sqrt{3mn}$ **32.** $23a^3b^4\sqrt{3ab}$ **33.** $13r^3s^2\sqrt{30rs}$
34. $15c^2d^5\sqrt{3cd}$

Page 11   The Pythagorean Theorem

1. 5   **2.** 13   **3.** 39   **4.** 65   **5.** 8   **6.** 24   **7.** 15
8. 20   **9.** 48   **10.** 16   **11.** 20   **12.** 9   **13.** no
14. yes   **15.** yes   **16.** no   **17.** no   **18.** yes   **19.** yes
20. yes   **21.** $5\sqrt{2}$ **22.** $3\sqrt{3}$ **23.** $4\sqrt{7}$ **24.** $5\sqrt{10}$
25. $6\sqrt{3}$ **26.** $3\sqrt{7}$ **27.** $7\sqrt{2}$ **28.** $8\sqrt{2}$

Page 12   The Pythagorean Theorem

1. 20   **2.** 68   **3.** 53   **4.** 26   **5.** 63   **6.** 80   **7.** 96
8. 60   **9.** 60   **10.** 48   **11.** 24   **12.** 13   **13.** no
14. yes   **15.** yes   **16.** no   **17.** yes   **18.** no   **19.** no
20. yes   **21.** $4\sqrt{5}$ **22.** $2\sqrt{7}$ **23.** $6\sqrt{11}$ **24.** $8\sqrt{5}$
25. $4\sqrt{13}$ **26.** $9\sqrt{3}$ **27.** $9\sqrt{3}$ **28.** $11\sqrt{5}$

Page 13   The Pythagorean Theorem

1. 65 m   **2.** 164 m   **3.** 159 cm   **4.** 145 cm   **5.** 119 km
6. 85 cm   **7.** yes   **8.** yes   **9.** $7\sqrt{11}$ m   **10.** $9\sqrt{7}$ cm
11. $8\sqrt{5}$ km   **12.** $9\sqrt{3}$ cm

Page 14   The Pythagorean Theorem

1. 85 km   **2.** 123 cm   **3.** 122 m   **4.** 73 cm   **5.** 148 m
6. 125 cm   **7.** no   **8.** no   **9.** $6\sqrt{7}$ cm   **10.** $10\sqrt{5}$ cm
11. $22\sqrt{2}$ m   **12.** $9\sqrt{7}$ cm

Page 15 Radical Expressions

1. $x + y$ 2. $p - q$ 3. $2a - b$ 4. $3c + 2d$ 5. $m + n$
6. $r - s$ 7. $3x - 2$ 8. $4x + 5$ 9. $5x + 2$ 10. $6x - 5$
11. $(x + 3)^2$ 12. $(x - 5)^3$ 13. $(5 - 7x)^4$
14. $(8 - 3x)^2$ 15. $(2x + 3)^3$ 16. $(4x - 7)\sqrt{4x - 7}$
17. $(2x - 3y)^4$ 18. $(4x - 9y)^3$ 19. $x^2 + 3x + 2$
20. $x^2 - 2x - 15$ 21. $8x^2 - 10x + 3$ 22. $6x^2 - 11x + 10$
23. $4(2x - 7)$ 24. $7(5x + 9)$ 25. $9(3x + 1)^2$
26. $10(2x - 13)^3$ 27. $4(8x - 1)\sqrt{3}$ 28. $6(10x - 3)\sqrt{2}$
29. $9x(3x + 1)\sqrt{2x(3x + 1)}$ 30. $7x^2(2x + 13)^2\sqrt{5x(2x + 13)}$
31. $6xy(5x - 7y)\sqrt{7y}$ 32. $18x^2(7x + 3y)\sqrt{2xy(7x + 3y)}$
33. $11x^2(10x + 3y)\sqrt{5y(10x + 3y)}$
34. $13y^3(5x - 17y)\sqrt{2x}$

Page 16 Radical Expressions

1. $m + n$ 2. $r - s$ 3. $2a - b$ 4. $3x + y$ 5. $x + 5$
6. $m - 4$ 7. $3x - 5$ 8. $4x + 3$ 9. $7x - 2$
10. $8x - 3$ 11. $(x - 7)^3$ 12. $(x + 9)^2$ 13. $(10 - 3x)^2$
14. $(7 - 13x)^4$ 15. $(4x + 17)\sqrt{4x + 17}$
16. $(3x - 19)^2\sqrt{3x - 19}$ 17. $(2x + 1)^3$ 18. $(3x - 7)^3$
19. $2x^2 + 11x + 5$ 20. $6x^2 - 5x - 4$ 21. $28x^2 + 25x - 3$
22. $12x^2 - 28x + 15$ 23. $5(3x + 5)$ 24. $6(7x - 3)$
25. $13(5 - 2x)^2$ 26. $15(10x^2 - 9x)$ 27. $5(3x + 8)\sqrt{7}$
28. $9(11 + 3x)\sqrt{5}$ 29. $3x^4(5x + 3)^2\sqrt{6x(5x + 3)}$
30. $9x^2(2x - 11)^2\sqrt{7x}$ 31. $5x^2y^2(3x - 7y)\sqrt{11y(3x - 7y)}$
32. $17x^3(2x + 9y)^3\sqrt{5y(2x + 9y)}$
33. $19x^3(8x + 3y)^2\sqrt{2y(8x + 3y)}$
34. $8x^3y^2(3x - 10y)\sqrt{(3x - 10y)}$

Page 17 Multiplication of Radical Expressions

1. $\sqrt{6}$ 2. 2 3. $\sqrt{30}$ 4. $4\sqrt{3}$ 5. 8 6. $2\sqrt{3}$ 7. $6\sqrt{5}$
8. $3\sqrt{10}$ 9. $3\sqrt{6}$ 10. 50 11. $15\sqrt{2}$ 12. $4\sqrt{3}$
13. $9\sqrt{10}$ 14. $28\sqrt{230}$ 15. $18\sqrt{10}$ 16. $3\sqrt{7}$
17. $4\sqrt{30}$ 18. $14\sqrt{15}$ 19. $3\sqrt{1518}$ 20. $26\sqrt{30}$
21. $9a^2b^3\sqrt{6}$ 22. $24xy^2\sqrt{xy}$ 23. $24c^3d^2\sqrt{154}$
24. $17pq^3\sqrt{6p}$ 25. $6r^4s^2\sqrt{30}$ 26. $24a^2b^2\sqrt{95b}$
27. $33x^2y^2\sqrt{3y}$ 28. $20m^2n^2\sqrt{6m}$ 29. $44a^3\sqrt{2a}$
30. $18x^3\sqrt{10x}$ 31. $680y^3\sqrt{3y}$ 32. $1728m^4$
33. $10r^5\sqrt{105r}$ 34. $50t^5\sqrt{6}$

Page 18 Multiplication of Radical Expressions

1. $2\sqrt{2}$ 2. $\sqrt{14}$ 3. $2\sqrt{14}$ 4. $\sqrt{10}$ 5. $2\sqrt{15}$
6. $6\sqrt{10}$ 7. 9 8. $2\sqrt{6}$ 9. $20\sqrt{30}$ 10. $5\sqrt{3}$
11. $4\sqrt{5}$ 12. $150\sqrt{21}$ 13. $4\sqrt{57}$ 14. $13\sqrt{6}$
15. $4\sqrt{33}$ 16. $8\sqrt{5}$ 17. $216\sqrt{10}$ 18. $240\sqrt{15}$
19. $8\sqrt{15}$ 20. $4\sqrt{1364}$ 21. $66xy\sqrt{6xy}$ 22. $4a^2b^3\sqrt{105}$
23. $4m^2n^2\sqrt{30mn}$ 24. $12r^2s^2\sqrt{14r}$ 25. $8x^2y^2\sqrt{30xy}$
26. $80a^4b^2\sqrt{154}$ 27. $22r^3s\sqrt{6s}$ 28. $3m^4n^2\sqrt{231n}$
29. $468a^4\sqrt{6a}$ 30. $54y^3\sqrt{2y}$ 31. $57y^4\sqrt{2}$ 32. $75m^4\sqrt{6}$
33. $90a^4\sqrt{3}$ 34. $567t^4\sqrt{34t}$

Page 19 Multiplication of Radical Expressions

1. 4 2. 7 3. $3\sqrt{2}$ 4. $6\sqrt{2}$ 5. 6 6. $\sqrt{21}$
7. $15\sqrt{2}$ 8. $7\sqrt{30}$ 9. $12\sqrt{3}$ 10. 6 11. $6\sqrt{13}$
12. $10\sqrt{30}$ 13. $4\sqrt{78}$ 14. $16\sqrt{5}$ 15. 9 16. $6\sqrt{30}$
17. 120 18. $90\sqrt{3}$ 19. $72\sqrt{110}$ 20. $384\sqrt{35}$
21. $20r^2s^3\sqrt{3r}$ 22. $3p^2q^2\sqrt{154p}$ 23. $54a^3b\sqrt{13b}$
24. $10r^2s^2\sqrt{17}$ 25. $8x^3y^3\sqrt{10x}$ 26. $20m^2n^3\sqrt{5m}$
27. $9a^2b^3\sqrt{11a}$ 28. $392x^4y^3\sqrt{3x}$ 29. $360r^4\sqrt{19}$
30. $42y^3\sqrt{7}$ 31. $18a^4\sqrt{42}$ 32. $60p^4\sqrt{6}$ 33. $208a^4\sqrt{a}$
34. $105z^3\sqrt{z}$

Page 20 Multiplication of Radical Expressions

1. 3 2. $\sqrt{42}$ 3. $4\sqrt{6}$ 4. $2\sqrt{6}$ 5. $3\sqrt{5}$ 6. $3\sqrt{5}$
7. 540 8. 16 9. $\sqrt{42}$ 10. $210\sqrt{15}$ 11. $4\sqrt{33}$
12. $9\sqrt{10}$ 13. $4\sqrt{13}$ 14. $8\sqrt{30}$ 15. $18\sqrt{11}$
16. $4\sqrt{85}$ 17. $24\sqrt{10}$ 18. $540\sqrt{6}$ 19. $8\sqrt{570}$
20. $60\sqrt{30}$ 21. $15p^2q^2\sqrt{6pq}$ 22. $4s^2t^2\sqrt{110s}$
23. $80x^3y^2\sqrt{22}$ 24. $70r^3s^2\sqrt{30}$ 25. $6a^3b^4\sqrt{55}$
26. $21m^4n^2\sqrt{6}$ 27. $420x^2y^2\sqrt{2x}$ 28. $19c^2d^3\sqrt{6}$
29. $21a^4\sqrt{11a}$ 30. $1656y^4$ 31. $12c^3\sqrt{42c}$
32. $78x^5\sqrt{2x}$ 33. $21r^4\sqrt{55}$ 34. $21z^3\sqrt{66z}$

Page 21 Multiplication of Radical Expressions

1. $3x\sqrt{2} + 2y\sqrt{2}$ 2. $2x\sqrt{3} - 5y\sqrt{3}$ 3. $4a\sqrt{7} - 9b\sqrt{7}$
4. $11m\sqrt{11} + 3n\sqrt{11}$ 5. $4x^2\sqrt{2} + 14x\sqrt{2} - 2\sqrt{2}$
6. $9x^2\sqrt{3} - 15x\sqrt{3} + 6\sqrt{3}$ 7. $5x^2\sqrt{3} - 25x\sqrt{3} + 15\sqrt{3}$
8. $2x^2\sqrt{3} - 18x\sqrt{3} - 2\sqrt{3}$ 9. $3x^3\sqrt{5} - 30x^2\sqrt{5} - 6x\sqrt{5}$
10. $4x^4\sqrt{3} - 28x^3\sqrt{3} + 12x^2\sqrt{3}$ 11. $9x^7\sqrt{2} + 63x^5\sqrt{2} -$
$45x^4\sqrt{2}$ 12. $3x^6\sqrt{7} - 6x^4\sqrt{7} + 15x^3\sqrt{7}$ 13. $3 + \sqrt{21}$
14. $\sqrt{35} + 5$ 15. $\sqrt{55} + 11$ 16. $13 + \sqrt{39}$ 17. $3\sqrt{5} + 5\sqrt{3}$
18. $2\sqrt{30} + 10$ 19. $2\sqrt{15} + 3\sqrt{10}$ 20. $7\sqrt{2} + 2\sqrt{35}$
21. $a^2\sqrt{10} - ab\sqrt{15}$ 22. $x^2\sqrt{21} + xy\sqrt{35}$
23. $m^2\sqrt{15} + mn\sqrt{21}$ 24. $zy\sqrt{6} - z^2\sqrt{14}$ 25. $2x^2\sqrt{7} -$
$6xy$ 26. $3m^2\sqrt{2} + 6mn$ 27. $7r^2\sqrt{2} + 7rs\sqrt{3}$
28. $11st\sqrt{2} - 11t^2\sqrt{3}$
29. $9x^2\sqrt{x} - 3xy\sqrt{6x}$ 30. $4x\sqrt{6y} - 12y\sqrt{y}$
31. $102m^4 - 4m^2n\sqrt{34mn}$ 32. $9a^2\sqrt{2} + 24b\sqrt{3ab}$
33. $100s^3\sqrt{2} + 8st\sqrt{10st}$ 34. $3x^2\sqrt{91} + 2y\sqrt{1365xy}$

Page 22 Multiplication of Radical Expressions

1. $7a\sqrt{5} + 3b\sqrt{5}$ 2. $5x\sqrt{2} - 4y\sqrt{2}$ 3. $5m\sqrt{13} - 2n\sqrt{13}$
4. $6r\sqrt{17} + 2t\sqrt{17}$ 5. $12x^2\sqrt{5} - 20x\sqrt{5} + 28\sqrt{5}$
6. $8x^2\sqrt{2} + 28x\sqrt{2} - 20\sqrt{2}$ 7. $25x^2\sqrt{5} - 15x\sqrt{5} + 5\sqrt{5}$
8. $9x^2\sqrt{3} - 63x\sqrt{3} - 27\sqrt{3}$ 9. $6x^3\sqrt{2} - 18x^2\sqrt{2} +$
$42x\sqrt{2}$ 10. $11x^4\sqrt{2} + 55x^3\sqrt{2} + 33x^2\sqrt{2}$ 11. $8x^5\sqrt{7} +$
$20x^4\sqrt{7} - 12x^3\sqrt{7}$ 12. $25x^6\sqrt{7} - 20x^5\sqrt{7} + 5x^4\sqrt{7}$
13. $\sqrt{10} + 2$ 14. $\sqrt{21} - 7$ 15. $\sqrt{55} - 5$ 16. $\sqrt{26} +$
13 17. $\sqrt{165} + 11\sqrt{3}$ 18. $3\sqrt{13} + 13\sqrt{3}$ 19. $2\sqrt{17} +$
$17\sqrt{2}$ 20. $5\sqrt{2} - 2\sqrt{5}$ 21. $x^2\sqrt{21} - xy\sqrt{35}$
22. $bc\sqrt{10} + c^2\sqrt{15}$ 23. $a^2\sqrt{14} + ab\sqrt{6}$ 24. $xy\sqrt{22} -$
$y^2\sqrt{55}$ 25. $3x^2\sqrt{5} - 3xy\sqrt{10}$ 26. $x^2\sqrt{21} - 7xy\sqrt{2}$
27. $5st\sqrt{2} + 5t^2\sqrt{3}$ 28. $13v^2\sqrt{2} - 13vw\sqrt{3}$
29. $4x\sqrt{22x} - 4y\sqrt{33x}$ 30. $3y^2\sqrt{15y} - 15y\sqrt{y}$
31. $32a^2\sqrt{6a} - 80ab\sqrt{5b}$ 32. $54c^2\sqrt{3c} - 126cd\sqrt{d}$
33. $10r^4\sqrt{15} + 60r^2s\sqrt{2rs}$ 34. $60p^2 - 32q\sqrt{6pq}$

Page 23 Division of Rational Expressions

1. 4 2. 7 3. 9 4. 8 5. $\sqrt{2}/2$ 6. $\sqrt{6}/3$
7. $(2\sqrt{5})/5$ 8. $(3\sqrt{10})/10$ 9. $(4\sqrt{3})/3$ 10. $(8\sqrt{7})/7$
11. $(7\sqrt{2})/2$ 12. $(9\sqrt{5})/5$ 13. $\sqrt{3}$ 14. $\sqrt{13}$ 15. $\sqrt{5}$
16. $\sqrt{6}$ 17. $\sqrt{105}/7$ 18. $\sqrt{105}/5$ 19. $\sqrt{22}/2$
20. $\sqrt{51}/3$ 21. $\sqrt{42}/3$ 22. $(3\sqrt{5})/5$ 23. $(2\sqrt{42})/7$
24. $(5\sqrt{7})/7$ 25. $(5\sqrt{11})/11$ 26. $(9\sqrt{17})/17$

Page 24 Division of Rational Expressions

1. 7 2. 12 3. 4 4. 11 5. $(3\sqrt{5})/5$ 6. $(5\sqrt{7})/7$
7. $\sqrt{14}/4$ 8. $\sqrt{30}/10$ 9. $(10\sqrt{7})/7$ 10. $(12\sqrt{5})/5$
11. $(11\sqrt{13})/13$ 12. $(13\sqrt{11})/11$ 13. 2 14. $\sqrt{3}$
15. $\sqrt{19}$ 16. $\sqrt{10}$ 17. $\sqrt{330}/10$ 18. $\sqrt{2310}/55$
19. $\sqrt{87}/3$ 20. $\sqrt{646}/17$ 21. $(2\sqrt{154})/11$
22. $\sqrt{385}/7$ 23. $(2\sqrt{285})/19$ 24. $(9\sqrt{2})/2$
25. $(7\sqrt{2})/2$ 26. $(11\sqrt{2})/2$

Page 25 Division of Rational Expressions

1. $\sqrt{2}/2$ 2. $\sqrt{13}/13$ 3. $(31\sqrt{15})/15$ 4. $(38\sqrt{7})/7$
5. $(3\sqrt{15})/5$ 6. $\sqrt{3}/3$ 7. $\sqrt{30}/4$ 8. $(5\sqrt{39})/13$
9. $\sqrt{5}/5$ 10. $\sqrt{210}/12$ 11. $\sqrt{87}/3$ 12. $(3\sqrt{35})/7$
13. $\sqrt{114}/18$ 14. $\sqrt{39}/9$ 15. $(17\sqrt{15})/25$ 16. $\sqrt{2}/4$
17. $(cd\sqrt{5e})/e$ 18. $(b\sqrt{7abc})/c$ 19. $\sqrt{x}/x$ 20. $\sqrt{3y}/y$
21. $\sqrt{11y}/y^2$ 22. $\sqrt{x}/x^2$ 23. $(2\sqrt{pq})/q$ 24. $(3\sqrt{de})/e$
25. $(7\sqrt{6mn})/3n^2$ 26. $(6x\sqrt{10y})/5y^2$

Page 26 Division of Rational Expressions

1. $\sqrt{13}/13$ 2. $\sqrt{7}/7$ 3. $(36\sqrt{11})/11$ 4. $(39\sqrt{5})/5$
5. $(8\sqrt{6})/3$ 6. $\sqrt{7}/49$ 7. $(5\sqrt{35})/7$ 8. $\sqrt{42}/4$
9. $\sqrt{6}/6$ 10. $(4\sqrt{5})/5$ 11. $\sqrt{598}/13$ 12. $(9\sqrt{15})/5$
13. $\sqrt{91}/28$ 14. $\sqrt{105}/35$ 15. $13/2$ 16. $1/27$
17. $\sqrt{14art}/t$ 18. $(2a\sqrt{3bc})/b$ 19. $\sqrt{z}/z$ 20. $\sqrt{5q}/q$
21. $\sqrt{7m}/m^2$ 22. $\sqrt{10y}/y^2$ 23. $(2\sqrt{33yz})/3y$
24. $(5\sqrt{rt})/t$ 25. $(3\sqrt{55a})/5b^2$ 26. $(2t\sqrt{42r})/3r^3$

Page 27 Division of Rational Expressions

1. $6 - 7\sqrt{2}$ 2. $5 + 8\sqrt{5}$ 3. $1 + 8\sqrt{3}$ 4. $3 + \sqrt{2}$
5. $2 + 3\sqrt{6}$ 6. $12 - 5\sqrt{15}$ 7. -41 8. -13
9. $(8\sqrt{3} - 3)/3$ 10. $(20 + 10\sqrt{5})/5$ 11. $(7 + 9\sqrt{7})/7$
12. $6\sqrt{2} - 1$ 13. $(\sqrt{6} + 2\sqrt{3})/2$ 14. $(5 - \sqrt{10})/5$
15. $(3\sqrt{35} - 10)/5$ 16. $(8\sqrt{33} + 9)/3$
17. $(4\sqrt{10} + 9\sqrt{15})/5$ 18. $(28\sqrt{21} - 30\sqrt{15})/3$
19. $(20\sqrt{21} - 6\sqrt{42})/7$ 20. $9\sqrt{14} + 14\sqrt{10}$ 21. $4\sqrt{6} + 42$
22. $(25\sqrt{35} + 12\sqrt{15})/5$ 23. $(52 - 9\sqrt{130})/13$
24. $63 - 25\sqrt{3}$ 25. $(252\sqrt{14} + 169\sqrt{35})/7$
26. $(20\sqrt{33} + 63)/3$

Page 28 Division of Rational Expressions

1. $8 - 3\sqrt{7}$ 2. $9 + 3\sqrt{3}$ 3. $3\sqrt{2} + 2\sqrt{3}$
4. $9\sqrt{2} - 4\sqrt{3}$ 5. $21 + 6\sqrt{6}$ 6. $16 - 3\sqrt{5}$ 7. -4
8. $6 + \sqrt{11}$ 9. $(2\sqrt{10} + 5)/5$ 10. $(105 + 10\sqrt{7})/7$
11. $(143 + 3\sqrt{11})/11$ 12. $(3\sqrt{6} - 6)/2$ 13. $(\sqrt{10} + 2\sqrt{5})/2$
14. $(7\sqrt{2} + \sqrt{105})/7$ 15. $4\sqrt{5} + \sqrt{6}$
16. $(39\sqrt{2} + \sqrt{65})/13$ 17. $(21\sqrt{6} + 16\sqrt{10})/2$
18. $(20\sqrt{6} - 6\sqrt{21})/3$ 19. $(15\sqrt{35} - 49\sqrt{14})/7$
20. $(28\sqrt{77} + 18\sqrt{55})/11$ 21. $(10\sqrt{10} + 27)/2$
22. $(24\sqrt{105} + 30\sqrt{14})/7$ 23. $(120\sqrt{3} - 48\sqrt{15})/5$
24. $(27\sqrt{2} - 28\sqrt{21})/9$ 25. $(8\sqrt{22} + 9\sqrt{26})/2$
26. $(8\sqrt{21} - 27\sqrt{5})/3$

Page 29 Factoring Radical Expressions

1. $\sqrt{5}(7x + 2y)$ 2. $\sqrt{2}(3c + 5d)$ 3. $\sqrt{3}(11y - 7z)$
4. $\sqrt{11}(4t - 7s)$ 5. $2\sqrt{3b}(a + 2c)$ 6. $3\sqrt{5y}(2x - 3z)$
7. $4\sqrt{5r}(3t - 2s)$ 8. $2\sqrt{3z}(3z + 7y)$
9. $\sqrt{2}(x^2 - 7x + 1)$ 10. $\sqrt{3}(x^2 - 5x + 2)$
11. $\sqrt{5}(x^2 + 9x + 3)$ 12. $\sqrt{7}(x^2 + 2x - 9)$
13. $\sqrt{3}(3x + 2y)$ 14. $\sqrt{5}(3a - 4b)$
15. $\sqrt{2}(8r + 5t)$ 16. $\sqrt{3}(7c - 9d)$
17. $q\sqrt{5}(3p - 5q)$ 18. $x\sqrt{7}(4b - 7a)$
19. $t\sqrt{7t}(2c + 3d)$ 20. $b\sqrt{5}(4x + 7y)$
21. $x\sqrt{11}(32x + 35y)$
22. $t\sqrt{2}(190r - 33s)$ 23. $21d\sqrt{5}(c - 3d)$
24. $t\sqrt{6}(8t + 35z)$ 25. $\sqrt{2}(2x^2 - 12x + 5)$
26. $\sqrt{3}(3x^2 - 6x + 5)$ 27. $\sqrt{3}(3x^2 - 7x - 2)$
28. $\sqrt{3}(5x^2 - 3x + 1)$ 29. $\sqrt{3}(x + y)(x - y)$
30. $\sqrt{2}(2a + 3b)(2a - 3b)$ 31. $\sqrt{5}(x + y)^2$
32. $\sqrt{7}(x + 2y)(x + y)$ 33. $\sqrt{3}(x + 3y)(x + 2y)$
34. $\sqrt{2}(2y + x)^2$

Page 30 Factoring Radical Expressions

1. $\sqrt{2}(5a + 7b)$ 2. $5\sqrt{3}(x + 2y)$ 3. $\sqrt{5}(4m - 3n)$
4. $\sqrt{7}(2r + 9s)$ 5. $\sqrt{3x}(5x + 2y)$ 6. $4\sqrt{2b}(2a - b)$
7. $3\sqrt{7b}(a - 2b)$ 8. $2\sqrt{3m}(5m - 6n)$
9. $\sqrt{11}(x^2 + 8x + 2)$ 10. $\sqrt{7}(x^2 - 7x - 3)$
11. $\sqrt{5}(x^2 + 5x + 10)$ 12. $\sqrt{13}(x^2 - 3x + 1)$
13. $5\sqrt{2}(a + 9)$ 14. $\sqrt{3}(9c - 4d)$ 15. $\sqrt{5}(4t - 9s)$
16. $\sqrt{7}(6x + 11y)$ 17. $n\sqrt{3}(5m - 4n)$
18. $5t\sqrt{2}(3r - 4t)$ 19. $q\sqrt{3q}(13p + 10)$
20. $\sqrt{2d}(8c + 21d^2)$
21. $10a\sqrt{11}(3a - 7b)$ 22. $3m\sqrt{7}(12m - 5n)$
23. $14t\sqrt{3}(5t - 6v)$ 24. $8y\sqrt{17}(5x - 12y)$
25. $\sqrt{3}(2x^2 - 9x + 9)$ 26. $\sqrt{2}(6x^2 - 8x - 7)$
27. $\sqrt{5}(4x^2 - 20x - 9)$ 28. $\sqrt{7}(4x^2 - 42x + 9)$
29. $\sqrt{11}(a + b)(a - b)$ 30. $9\sqrt{3}(a + 2b)(a - 2b)$
31. $\sqrt{7}(a + 6)(a - 1)$ 32. $\sqrt{5}(x + 3)(x + 4)$
33. $\sqrt{3}(a - 6)(a - 1)$ 34. $\sqrt{2}(a - 4)(a + 1)$

Page 31 Factoring Radical Expressions

1. $4\sqrt{2}$ 2. $7\sqrt{3}$ 3. $-3\sqrt{7}$ 4. $5\sqrt{5}$ 5. $2\sqrt{5}$
6. $-7\sqrt{11}$ 7. 0 8. $7\sqrt{15}$ 9. $-2\sqrt{3}$ 10. 0 11. $6\sqrt{7}$
12. $-8\sqrt{11}$ 13. $7\sqrt{3} - 6\sqrt{2}$ 14. $7\sqrt{3} - \sqrt{2}$
15. $9\sqrt{10} - 2\sqrt{5}$ 16. $-3\sqrt{6} - 3\sqrt{2}$ 17. $-8\sqrt{2}$
18. $15\sqrt{3}$ 19. $5\sqrt{5}$ 20. $-23\sqrt{7}$
21. $(13\sqrt{6} - 36\sqrt{3})/36$ 22. $(3\sqrt{10} - 4\sqrt{14})/12$
23. $-\sqrt{21}/6$ 24. $(12\sqrt{15} - 5\sqrt{35})/50$
25. $(11\sqrt{30} + 5\sqrt{21})/30$ 26. $(59\sqrt{10})/70$
27. $(-51\sqrt{14})/182$ 28. $(-17\sqrt{15})/60$

Page 32 Addition and Subtraction of Radical Expressions

1. $-4\sqrt{5}$ 2. $10\sqrt{7}$ 3. $16\sqrt{3}$ 4. $6\sqrt{2}$ 5. $7\sqrt{2}$ 6. 0
7. $-2\sqrt{15}$ 8. $9\sqrt{7}$ 9. $10\sqrt{13}$ 10. $12\sqrt{3}$ 11. 0
12. $2\sqrt{10}$ 13. $3\sqrt{2} - 14\sqrt{3}$ 14. $12\sqrt{7} - 4\sqrt{10}$
15. $5\sqrt{13} - 7\sqrt{19}$ 16. $31\sqrt{21} + 3\sqrt{17}$ 17. $111\sqrt{3}$
18. $-26\sqrt{2}$ 19. $66 - 25\sqrt{11}$ 20. $66\sqrt{10}$
21. $(131\sqrt{15})/60$ 22. $(-97\sqrt{10})/70$ 23. $29\sqrt{6}/30$
24. $(8\sqrt{14})/7$ 25. $(23\sqrt{21})/105$
26. $(4\sqrt{15} - 23\sqrt{30})/30$ 27. $5\sqrt{6}$ 28. $\sqrt{10}/10$

Page 33 Addition and Subtraction of Radical Expressions

1. $25ab\sqrt{11}$ 2. $-81cd\sqrt{6}$ 3. $-58cd\sqrt{7}$ 4. $219rs\sqrt{2}$
5. $-65rs\sqrt{13}$ 6. $-39mn\sqrt{3}$ 7. $269ab\sqrt{10}$
8. $-110rs\sqrt{5}$ 9. $481a^2\sqrt{3a}$ 10. $241ab\sqrt{7b}$
11. $25yz\sqrt{6z}$ 12. $39xy\sqrt{5x}$ 13. $192a^2b^2\sqrt{10b}$
14. $-90c^2d^2\sqrt{13c}$ 15. $-54m^2n^3\sqrt{2m}$ 16. $276r^2s^2\sqrt{11r}$
17. $142xy\sqrt{3xy}$ 18. $-147s^2t^2\sqrt{7st}$ 19. $-40m^2n^2\sqrt{10mn}$
20. $345y^2z^2\sqrt{5yz}$ 21. $(9b\sqrt{35a})/7a$ 22. $(-2r\sqrt{35s})/5s$
23. $(-2x\sqrt{30y})/y$ 24. $(28a^2\sqrt{22b})/11b$
25. $(27\sqrt{65ab})/65b$ 26. $(-25x\sqrt{21y})/21y^2$
27. $(74a\sqrt{66ab})/66b$ 28. $(67x\sqrt{22y})/22y^2$

Page 34 Addition and Subtraction of Radical Expressions

1. $58xy\sqrt{10}$ 2. $-ab\sqrt{7}$ 3. $125cd\sqrt{5}$ 4. $-64rs\sqrt{6}$
5. $-78pq\sqrt{13}$ 6. $246mn\sqrt{2}$ 7. $211vw\sqrt{3}$
8. $212pq\sqrt{11}$ 9. $-36a^2\sqrt{5b}$ 10. $-104x^2y\sqrt{2x}$
11. $259c^2d\sqrt{11c}$ 12. $86b^2c^2\sqrt{10c}$ 13. $-66x^2y^2\sqrt{3y}$
14. $332x^2y^2\sqrt{13x}$ 15. $-57c^2d^2\sqrt{6d}$ 16. $-65a^2b\sqrt{7b}$
17. $184c^2d^2\sqrt{2cd}$ 18. $-112x^2y^2\sqrt{11xy}$
19. $294a^2b^2\sqrt{13ab}$ 20. $285p^2q^2\sqrt{3pq}$
21. $(-35a\sqrt{22b})/22b$ 22. $(121\sqrt{78xy})/78y$
23. $(101p\sqrt{30q})/30q$ 24. $(17\sqrt{10mn})/10n$
25. $(-26\sqrt{35rs})/35s$ 26. $(2x\sqrt{21y})/21y$
27. $(3v\sqrt{130w})/5w$ 28. $(97a\sqrt{66b})/66b$

Page 35 Multipication of Binomials Containing Radicals

1. $3 + \sqrt{6} + \sqrt{15} + \sqrt{10}$ 2. $\sqrt{70} + 7\sqrt{2} + 2\sqrt{10} + 2\sqrt{14}$
3. $\sqrt{22} + \sqrt{77} - \sqrt{10} - \sqrt{35}$ 4. $12\sqrt{2} + 4\sqrt{6} + 2\sqrt{3} + 2$
5. $6\sqrt{26} - 3\sqrt{65} + 8 - 2\sqrt{10}$ 6. $5\sqrt{2} + \sqrt{10} - \sqrt{30} - \sqrt{6}$
7. 9 8. $2\sqrt{65} + \sqrt{26} - 10 - \sqrt{10}$
9. $3\sqrt{2} - \sqrt{15} + 4\sqrt{3} - 2\sqrt{10}$ 10. $3\sqrt{14} + 3\sqrt{10} + \sqrt{35} + 5$
11. $\sqrt{15} + \sqrt{30} - \sqrt{6} - 2\sqrt{3}$ 12. $49\sqrt{2} + 49 - \sqrt{70} - \sqrt{35}$
13. $5\sqrt{10} + 5\sqrt{6} - \sqrt{30} - 3\sqrt{2}$
14. $10\sqrt{6} - 2\sqrt{105} + \sqrt{30} - \sqrt{21}$
15. $2\sqrt{22} - 2\sqrt{14} - \sqrt{55} + \sqrt{35}$
16. $14\sqrt{30} + 2\sqrt{5} + 14\sqrt{6} + 4$ 17. $13 - \sqrt{39} - \sqrt{91} - \sqrt{21}$
18. $\sqrt{39} + \sqrt{21} - \sqrt{78} - \sqrt{42}$
19. $7\sqrt{22} - 7\sqrt{55} - 4 - 2\sqrt{10}$ 20. 4
21. $31 + 4\sqrt{21}$ 22. $8 + 2\sqrt{15}$
23. $13 - 4\sqrt{10}$ 24. $35 + 12\sqrt{6}$ 25. $17 + 4\sqrt{15}$
26. $153 + 30\sqrt{2}$ 27. $16 + 2\sqrt{55}$ 28. $54 + 2\sqrt{35}$
29. 2 30. 7 31. 5 32. 13 33. 4 34. 21

Page 36 Multiplication of Binomials Containing Radicals

1. $\sqrt{10} + \sqrt{6} - \sqrt{15} - 3$ 2. $4\sqrt{5} + 2\sqrt{15} - 4 - 2\sqrt{3}$
3. $16\sqrt{3} - \sqrt{30} + 8\sqrt{10} - 5$ 4. $\sqrt{154} + 6\sqrt{21} - \sqrt{77} - 3\sqrt{42}$ 5. $14 + \sqrt{42} - 2\sqrt{7} - \sqrt{6}$ 6. $8\sqrt{6} + 2\sqrt{21} + 4\sqrt{10} + \sqrt{35}$ 7. $9\sqrt{2} - \sqrt{30} + 9 - \sqrt{15}$ 8. $\sqrt{143} + 5\sqrt{22} + \sqrt{78} + 10\sqrt{3}$ 9. $2\sqrt{42} + 7 - 6\sqrt{2} - \sqrt{21}$
10. $2\sqrt{65} - 4\sqrt{10} + \sqrt{78} - 4\sqrt{3}$ 11. $2\sqrt{30} + 12 + \sqrt{70} + 2\sqrt{21}$ 12. $2\sqrt{30} + 2\sqrt{6} - 9\sqrt{5} - 9$ 13. $\sqrt{143} - 3\sqrt{39} + \sqrt{22} - 3\sqrt{6}$ 14. $\sqrt{33} + \sqrt{15} - 4\sqrt{77} - 4\sqrt{35}$
15. $2\sqrt{21} + \sqrt{42} - 2\sqrt{6} - 2\sqrt{3}$ 16. -5 17. $4\sqrt{15} + 2\sqrt{30} + \sqrt{42} + \sqrt{21}$ 18. $4\sqrt{22} + 16 - \sqrt{66} - 4\sqrt{3}$
19. $4 - 10\sqrt{6} + \sqrt{14} - 5\sqrt{21}$ 20. $\sqrt{78} - 2\sqrt{91} - 6 - 2\sqrt{42}$ 21. $12 + 4\sqrt{5}$ 22. $17 + 4\sqrt{15}$ 23. $14 + 2\sqrt{33}$
24. $15 - 10\sqrt{2}$ 25. $66 + 36\sqrt{2}$ 26. $32 + 8\sqrt{15}$
27. $22 + 4\sqrt{10}$ 28. $37 - 2\sqrt{210}$ 29. -7 30. 8
31. 5 32. -13 33. 13 34. 8

Page 37 Multiplication of Binomials Containing Radicals

1. $2\sqrt{33} + 6 - \sqrt{55} - \sqrt{15}$ 2. $\sqrt{102} - 2\sqrt{3} + \sqrt{34} - 2$
3. $2\sqrt{6} + 2\sqrt{10} - 3 - \sqrt{15}$ 4. $\sqrt{22} - \sqrt{66} - \sqrt{6} + 3\sqrt{2}$
5. $2\sqrt{21} - 6\sqrt{3} - 10\sqrt{7} + 30$ 6. $5 + 2\sqrt{30} - \sqrt{35} - 2\sqrt{42}$ 7. $\sqrt{30} + 2\sqrt{70} + \sqrt{15} + 2\sqrt{35}$ 8. $\sqrt{70} - 3\sqrt{30} + \sqrt{21} - 9$ 9. $1 + 2\sqrt{14}$ 10. $\sqrt{143} - \sqrt{26} - \sqrt{66} + 2\sqrt{3}$ 11. $2\sqrt{14} - 3\sqrt{35} - 8 + 6\sqrt{10}$ 12. $\sqrt{70} + 2\sqrt{15} - \sqrt{35} - \sqrt{30}$ 13. $\sqrt{77} + \sqrt{33} - 14 - 2\sqrt{21}$ 14. $6\sqrt{15} + 12\sqrt{3} - \sqrt{70} - 2\sqrt{14}$ 15. $12 + 2\sqrt{21} - 2\sqrt{33} - \sqrt{77}$
16. $\sqrt{182} + \sqrt{42} - 2\sqrt{78} - 6\sqrt{2}$ 17. $\sqrt{182} - 2\sqrt{26} - 14\sqrt{2} + 4\sqrt{14}$ 18. $\sqrt{78} - \sqrt{30} - 2\sqrt{26} + 2\sqrt{10}$
19. $12 + 2\sqrt{15} + 2\sqrt{6} + \sqrt{10}$ 20. $\sqrt{210} - 2\sqrt{7} + 9\sqrt{5} - 3\sqrt{6}$
21. $18 + 2\sqrt{65}$ 22. $17 - 2\sqrt{66}$ 23. $21 - 6\sqrt{6}$
24. $26 + 4\sqrt{30}$ 25. $16 + 2\sqrt{55}$ 26. $23 - 6\sqrt{10}$
27. $19 - 4\sqrt{21}$ 28. $32 + 4\sqrt{15}$ 29. -2 30. 8
31. 6 32. -4 33. 6 34. 12

Page 38 Multiplication of Binomials Containing Radicals

1. $\sqrt{21} + 2\sqrt{6} - \sqrt{14} - 4$ 2. $\sqrt{30} + 12 - \sqrt{10} + 4\sqrt{3}$
3. $2\sqrt{15} - 5 + 4\sqrt{6} - 2\sqrt{10}$ 4. $\sqrt{21} + \sqrt{35} - 2\sqrt{6} - 2\sqrt{10}$ 5. 1 6. $\sqrt{30} - 2\sqrt{35} - \sqrt{42} + 14$ 7. $\sqrt{55} - 2\sqrt{22} + \sqrt{10} - 4$ 8. $4\sqrt{5} + \sqrt{30} + 4 + \sqrt{6}$ 9. $2\sqrt{39} + 12\sqrt{2} + \sqrt{78} + 12$ 10. $6\sqrt{2} - \sqrt{42} - 6 + \sqrt{21}$
11. $\sqrt{130} - \sqrt{78} - \sqrt{30} + 3\sqrt{2}$ 12. $2\sqrt{22} + 2\sqrt{6} - \sqrt{33} - 3$ 13. $6 - 4\sqrt{6} - \sqrt{15} - 2\sqrt{10}$ 14. $\sqrt{42} - 2\sqrt{14} - 2\sqrt{3} + 4$ 15. $2\sqrt{14} - 2\sqrt{21} + 2\sqrt{6} - 6$
16. $4\sqrt{6} + 2\sqrt{10} + 2\sqrt{15} + 5$ 17. $2\sqrt{15} + 4\sqrt{5} - 6 + 4\sqrt{3}$
18. $2\sqrt{14} - 7 - 2\sqrt{6} + \sqrt{21}$ 19. $3\sqrt{21} - 21 + 3\sqrt{2} - \sqrt{42}$
20. $2\sqrt{14} + 2\sqrt{6} - \sqrt{42} - 3\sqrt{2}$ 21. $15 + 2\sqrt{26}$
22. $80 - 10\sqrt{15}$ 23. $39 - 4\sqrt{77}$ 24. $13 + 4\sqrt{10}$

Page 39 Fractions Containing Radicals

1. $-(\sqrt{2} - \sqrt{3})$ 2. $(4\sqrt{2} - 2\sqrt{5})/3$ 3. $(\sqrt{10} - \sqrt{2})/2$
4. $-2\sqrt{2} - 2\sqrt{5}$ 5. $-(3\sqrt{3} + 3\sqrt{5})/2$ 6. $-9\sqrt{5} - 9\sqrt{6}$
7. $-\sqrt{2} - \sqrt{6}$ 8. $(12\sqrt{3} + 6\sqrt{2})/5$
9. $(-29\sqrt{31} - 87)/22$ 10. $(5\sqrt{13} + 5\sqrt{2})/11$
11. $(-17\sqrt{14} + 51\sqrt{2})/4$ 12. $(5\sqrt{17} - 5\sqrt{11})/2$
13. $-15\sqrt{6} - 15\sqrt{7}$ 14. $(4\sqrt{17} + 4\sqrt{3})/7$
15. $(-7\sqrt{3} - 7\sqrt{7})/4$ 16. $(15\sqrt{2} + 10\sqrt{3})/3$
17. $-16 + 8\sqrt{5}$
18. $(7\sqrt{19} - 7\sqrt{3})/8$ 19. $(5\sqrt{13} + 5\sqrt{5})/4$
20. $(9\sqrt{17} + 9\sqrt{3})/7$ 21. $(20\sqrt{3} + 10\sqrt{6})/3$
22. $(69\sqrt{2} + 23\sqrt{5})/13$ 23. $(25\sqrt{14} - 25\sqrt{5})/9$
24. $(54\sqrt{5} + 27\sqrt{3})/17$ 25. $(26\sqrt{19} - 26\sqrt{2})/17$
26. $(22\sqrt{31} + 44\sqrt{2})/23$

Page 40 Fractions Containing Radicals

1. $4\sqrt{3} + 2\sqrt{5}$ 2. $-\sqrt{2} + \sqrt{6}$ 3. $(2\sqrt{2} + \sqrt{3})/5$
4. $-2\sqrt{3} + 5$ 5. $(5\sqrt{10} - 5\sqrt{3})/7$
6. $(-17\sqrt{2} + 17\sqrt{7})/5$ 7. $(3\sqrt{14} + 3\sqrt{2})/2$
8. $(-2\sqrt{3} - 2\sqrt{6})/3$ 9. $(7\sqrt{15} - 21)/6$
10. $(-11\sqrt{7} + 88)/54$
11. $(9\sqrt{14} - 18\sqrt{3})/2$ 12. $(6\sqrt{5} + 3\sqrt{7})/13$
13. $(5\sqrt{15} - 5\sqrt{2})/13$ 14. $(-22\sqrt{10} - 110)/15$
15. $3\sqrt{5} - 3$ 16. $2\sqrt{3} - \sqrt{6}$ 17. $-4\sqrt{7} - 16$
18. $69 + 46\sqrt{2}$ 19. $(-9\sqrt{14} + 18\sqrt{6})/5$
20. $(5\sqrt{6} - 5\sqrt{2})/4$ 21. $(-8\sqrt{7} + 32)/3$
22. $(247 - 19\sqrt{3})/163$ 23. $(30\sqrt{3} + 15\sqrt{5})/7$
24. $(25\sqrt{17} + 25\sqrt{5})/12$ 25. $(-9\sqrt{7} - 9\sqrt{3})/2$
26. $3\sqrt{14} + 3\sqrt{7}$

Page 41 Simple Radical Equations

1. $x = 49$ 2. $x = 9$ 3. $x = 225$ 4. $x = 169$ 5. $x = 12$
6. $x = 50$ 7. $x = 294$ 8. $x = 90$ 9. $x = -2$ 10. $x = 13$ 11. $x = 127$ 12. $x = -4$ 13. $x = 11$ 14. $x = -13$ 15. $x = -4$ 16. $x = -9/4$ 17. $x = 6$ 18. $x = -22$ 19. $x = -5$ 20. $x = 28$ 21. $x = 7 - 4\sqrt{3}$ 22. ϕ
23. $x = 69 - 28\sqrt{5}$ 24. $x = 19 + 4\sqrt{22}$ 25. ϕ
26. $x = 56 + 6\sqrt{55}$ 27. $x = 68 + 12\sqrt{21}$
28. $x = 194 - 48\sqrt{15}$ 29. $x = (13 - 2\sqrt{30})/3$
30. $x = (13 + 6\sqrt{5})/7$ 31. $x = (24 - 6\sqrt{14})/5$
32. $x = (25 - 6\sqrt{15})/2$ 33. $x = (25 + 8\sqrt{3})/4$ 34. ϕ

Page 42 Simple Radical Equations

1. $x = 100$ 2. $x = 16$ 3. $x = 625$ 4. $x = 49$ 5. $x = 45$ 6. $x = 128$ 7. $x = 176$ 8. $x = 300$ 9. $x = 76$
10. $x = -1$ 11. $x = 11$ 12. $x = 28$ 13. $x = 22$
14. $x = 3$ 15. $x = 3$ 16. $x = 30$ 17. $x = -12$
18. $x = 12$ 19. $x = 117$ 20. $x = 8$ 21. $x = 29 + 10\sqrt{2}$
22. ϕ 23. $x = 43 - 4\sqrt{30}$ 24. $x = 45 + 20\sqrt{2}$
25. $x = 36 + 4\sqrt{21}$ 26. $x = 48 - 12\sqrt{15}$ 27. ϕ
28. $x = 74 + 36\sqrt{2}$ 29. ϕ 30. $x = -83 - 12\sqrt{55}$
31. $x = (9 + 2\sqrt{14})/2$ 32. $x = (-22 - 9\sqrt{5})/3$
33. $x = -12 - 2\sqrt{55}$ 34. ϕ

Page 43 The Distance Formula

1. 20 2. 52 3. 37 4. 25 5. 17 6. 65 7. 25
8. 39 9. 29 10. 17 11. 10 12. 29 13. 17
14. 26 15. 29 16. 37 17. $\sqrt{89}$ 18. $\sqrt{410}$
19. $3\sqrt{10}$ 20. $12\sqrt{2}$ 21. $\sqrt{394}$ 22. $\sqrt{1009}$
23. $2\sqrt{65}$ 24. $2\sqrt{145}$ 25. $5\sqrt{5}$ 26. $\sqrt{493}$ 27. $\sqrt{353}$
28. $14\sqrt{5}$ 29. $3\sqrt{61}$ 30. $\sqrt{1202}$

Page 44 The Distance Formula

1. 17 **2.** 39 **3.** 26 **4.** 20 **5.** 37 **6.** 17 **7.** 37
8. 65 **9.** 5 **10.** 29 **11.** 52 **12.** 17 **13.** 25 **14.** 25
15. 29 **16.** 30 **17.** $3\sqrt{10}$ **18.** $\sqrt{685}$ **19.** $\sqrt{65}$
20. $\sqrt{922}$ **21.** $9\sqrt{2}$ **22.** $\sqrt{986}$ **23.** $2\sqrt{10}$ **24.** $\sqrt{481}$
25. $6\sqrt{5}$ **26.** $15\sqrt{2}$ **27.** $2\sqrt{85}$ **28.** $\sqrt{818}$ **29.** $2\sqrt{370}$
30. $\sqrt{85}$

Page 45 Simple Quadratic Equations

1. $x = \pm 4$ **2.** $x = \pm 3$ **3.** $x = \pm 7$ **4.** $x = \pm 9$ **5.** $x = \pm 6$ **6.** $x = \pm 8$ **7.** $x = \pm 10$ **8.** $x = \pm 11$ **9.** $x = \pm 3$
10. $x = \pm 5$ **11.** $x = \pm 26$ **12.** $x = \pm 2$ **13.** $x = \pm 16$
14. $x = \pm 17$ **15.** $x = \pm 13$ **16.** $x = \pm 18$ **17.** $x = \pm 25$
18. $x = \pm 14$ **19.** $x = \pm 15$ **20.** $x = \pm 19$ **21.** $x = \pm 5/4$
22. $x = \pm 3/7$ **23.** $x = \pm 2/3$ **24.** $x = \pm 11/6$ **25.** $x = \pm\sqrt{7}/5$ **26.** $x = \pm\sqrt{10}/9$ **27.** $x = \pm\sqrt{14}/13$ **28.** $x = \pm\sqrt{17}/15$ **29.** $x = \pm\sqrt{6}/2$ **30.** $x = \pm\sqrt{35}/5$ **31.** $x = \pm(2\sqrt{14})/7$ **32.** $x = \pm(3\sqrt{33})/11$ **33.** $x = \pm(4\sqrt{165})/15$
34. $x = \pm(3\sqrt{35})/10$

Page 46 Simple Quadratic Equations

1. $x = \pm 10$ **2.** $x = \pm 18$ **3.** $x = \pm 2$ **4.** $x = \pm 15$
5. $x = \pm 5$ **6.** $x = \pm 6$ **7.** $x = \pm 3$ **8.** $x = \pm 9$ **9.** $x = \pm 12$
10. $x = \pm 14$ **11.** $x = \pm 7$ **12.** $x = \pm 4$ **13.** $x = \pm 30$
14. $x = \pm 11$ **15.** $x = \pm 5$ **16.** $x = \pm 16$ **17.** $x = \pm 17$
18. $x = \pm 8$ **19.** $x = \pm 50$ **20.** $x = \pm 13$ **21.** $x = \pm 13/6$
22. $x = \pm 10/11$ **23.** $x = \pm 7/5$ **24.** $x = \pm 4/15$
25. $x = \pm\sqrt{11}/4$ **26.** $x = \pm\sqrt{33}/9$ **27.** $x = \pm\sqrt{21}/8$
28. $x = \pm\sqrt{15}/14$ **29.** $x = \pm\sqrt{77}/7$ **30.** $x = \pm\sqrt{174}/6$
31. $x = \pm(13\sqrt{6})/3$ **32.** $x = \pm(7\sqrt{15})/5$
33. $x = \pm(9\sqrt{30})/10$ **34.** $x = \pm(11\sqrt{51})/17$

Page 47 Simple Quadratic Equations

1. $x = 2$ or -6 **2.** $x = 8$ or 2 **3.** $x = -5$ or -17 **4.** $x = -1$ or -15 **5.** $x = 23$ or 1 **6.** $x = 31$ or -1 **7.** $x = 30$ or -12 **8.** $x = -6$ or -34 **9.** $x = 9$ or -17 **10.** $x = 25$ or 1 **11.** $x = 22$ or -18 **12.** $x = 13$ or -25 **13.** $x = 24$ or -6 **14.** $x = 37$ or -15 **15.** $x = -3 + 4\sqrt{39}$ **16.** $x = -7 \pm 10\sqrt{3}$ **17.** $x = -2 \pm 5\sqrt{3}$ **18.** $x = 5 \pm 8\sqrt{6}$
19. $x = 9 \pm 14\sqrt{3}$ **20.** $x = -6 \pm 11\sqrt{11}$
21. $x = -7 \pm 18\sqrt{5}$
22. $x = -9 \pm 6\sqrt{7}$ **23.** $x = -4 \pm 7\sqrt{5}$ **24.** $x = 8 \pm 15\sqrt{6}$ **25.** $x = 1 \pm 12\sqrt{6}$ **26.** $x = -5 \pm 19\sqrt{3}$ **27.** $x = 4 \pm 16\sqrt{7}$ **28.** $x = 10 \pm 9\sqrt{11}$ **29.** $x = (-6 \pm 10\sqrt{3})/3$
30. $x = (35 \pm 26\sqrt{5})/5$ **31.** $x = (77 \pm 17\sqrt{11})/11$
32. $x = (6 \pm 13\sqrt{6})/6$ **33.** $x = (6 \pm 5\sqrt{5})/2$
34. $x = (-33 \pm 10\sqrt{3})/3$

Page 48 Simple Quadratic Equations

1. $x = 13$ or -7 **2.** $x = -3$ or -19 **3.** $x = -6$ or -10
4. $x = 18$ or -14 **5.** $x = 22$ or -4 **6.** $x = 18$ or 8
7. $x = 4$ or -14 **8.** ϕ **9.** $x = 8$ or -4 **10.** $x = 8$ or -22 **11.** $x = -2$ or -10 **12.** $x = 13$ or -15 **13.** $x = 18$ or -4 **14.** $x = -20$ or -16 **15.** ϕ **16.** $x = 21$ or -3
17. $x = -3 \pm 9\sqrt{7}$ **18.** $x = 3 \pm 14\sqrt{6}$ **19.** $x = 5 \pm 6\sqrt{2}$
20. $x = -4 \pm 16\sqrt{3}$ **21.** $x = 10 \pm 13\sqrt{3}$ **22.** $x = 2 \pm 4\sqrt{7}$ **23.** $x = -8 \pm 13\sqrt{2}$ **24.** ϕ **25.** $x = -2 \pm 5\sqrt{13}$
26. $x = -1 \pm 15\sqrt{3}$ **27.** ϕ **28.** $x = -3 \pm 6\sqrt{7}$
29. $x = (35 \pm 12\sqrt{5})/5$ **30.** $x = 1 \pm 3\sqrt{6}$ **31.** $x = (-6 \pm 7\sqrt{3})/3$ **32.** $x = (-56 \pm 25\sqrt{7})/7$ **33.** $x = (3 \pm 19\sqrt{3})/3$ **34.** $x = (18 \pm 11\sqrt{2})/6$

Page 49 Completing the Square

1. 4 **2.** 16 **3.** 100 **4.** 36 **5.** 169 **6.** 225 **7.** 81
8. 625 **9.** 81/4 **10.** 121/4 **11.** 961/4 **12.** 529/4

13. 441/4 **14.** 1089/4 **15.** 361/4 **16.** 2025/4
17. 1/36 **18.** 1/64 **19.** 1/256 **20.** 9/400
21. 9/64 **22.** 49/256 **23.** 81/400 **24.** 1/49
25. 9/100 **26.** 49/400 **27.** 16/49 **28.** 49/64
29. 4/81 **30.** 121/100 **31.** 25/196 **32.** 36/49
33. 64/49 **34.** 25/49

Page 50 Completing the Square

1. 1 **2.** 9 **3.** 49 **4.** 25 **5.** 144 **6.** 64 **7.** 121
8. 196 **9.** 1/4 **10.** 49/4 **11.** 169/4 **12.** 9/4
13. 625/4 **14.** 225/4 **15.** 289/4 **16.** 2601/4
17. 1/16 **18.** 1/100 **19.** 4/25 **20.** 1/196
21. 9/196 **22.** 1/9 **23.** 25/64 **24.** 1/400
25. 1/324 **26.** 1/25 **27.** 9/49 **28.** 25/256
29. 4/49 **30.** 49/81 **31.** 81/64 **32.** 9/256
33. 1/81 **34.** 169/100

Page 51 Quadratic Equations

1. $x = -1 \pm \sqrt{13}$ **2.** $x = -1 \pm \sqrt{22}$ **3.** $x = -1 \pm 2\sqrt{2}$
4. $x = -1 \pm 3\sqrt{2}$ **5.** $x = 2 \pm \sqrt{17}$ **6.** $x = 2 \pm \sqrt{29}$
7. $x = -2 \pm 2\sqrt{6}$ **8.** $x = -2 \pm 3\sqrt{3}$ **9.** $x = -3 \pm \sqrt{21}$
10. $x = -3 \pm \sqrt{23}$ **11.** $x = 3 \pm 4\sqrt{2}$ **12.** $x = 3 \pm 2\sqrt{10}$
13. $x = 4 \pm \sqrt{37}$ **14.** $x = 4 \pm \sqrt{29}$ **15.** $x = -4 \pm 3\sqrt{3}$
16. $x = -4 \pm 2\sqrt{6}$ **17.** $x = -6 \pm \sqrt{29}$
18. $x = -6 \pm 2\sqrt{6}$ **19.** $x = 10 \pm \sqrt{86}$ **20.** $x = 13$ or 1
21. $x = (-3 \pm \sqrt{10})/2$ **22.** $x = (-5 \pm 2\sqrt{7})/2$
23. $x = (-7 \pm \sqrt{42})/2$ **24.** $x = (-13 \pm 4\sqrt{10})/2$
25. $x = (-11 \pm \sqrt{173})/2$ **26.** $x = (-9 \pm \sqrt{33})/2$
27. $x = (5 \pm \sqrt{65})/2$ **28.** ϕ **29.** $x = (-3 \pm \sqrt{21})/2$
30. $x = (-1 \pm \sqrt{29})/2$ **31.** $x = (-5 \pm \sqrt{53})/2$
32. $-3 \pm \sqrt{14}$ **33.** ϕ **34.** 3 or 1

Page 52 Quadratic Equations

1. $x = -1 \pm \sqrt{6}$ **2.** $x = (5 \pm \sqrt{57})/2$ **3.** $x = 2 \pm \sqrt{7}$
4. $x = 4 \pm \sqrt{3}$ **5.** $x = (-3 \pm \sqrt{29})/2$ **6.** $x = -3 \pm \sqrt{14}$
7. ϕ **8.** $x = 1 \pm \sqrt{17}$ **9.** $x = 3 \pm \sqrt{22}$ **10.** $x = (7 \pm \sqrt{69})/2$ **11.** $x = -4 \pm \sqrt{21}$ **12.** ϕ **13.** $x = -2 \pm \sqrt{17}$
14. $x = (9 \pm \sqrt{33})/2$ **15.** $x = (7 \pm \sqrt{29})/2$ **16.** $x = 2 \pm \sqrt{13}$ **17.** $x = (9 \pm \sqrt{86})/2$ **18.** $x = -5 \pm \sqrt{13}$
19. $x = -1 \pm 2\sqrt{3}$ **20.** $x = (-5 \pm 4\sqrt{2})/2$ **21.** $x = 5 \pm 2\sqrt{2}$ **22.** $x = (-9 \pm \sqrt{69})/2$ **23.** $x = (11 \pm 2\sqrt{30})/2$
24. $x = 4 \pm \sqrt{15}$ **25.** $x = (-3 \pm \sqrt{93})/2$
26. $x = (7 \pm 2\sqrt{13})/2$ **27.** $x = (-5 \pm \sqrt{89})/2$
28. $x = 1 \pm 3\sqrt{2}$ **29.** $x = -6 \pm 3\sqrt{3}$ **30.** $x = 3 \pm \sqrt{22}$
31. $x = 2 \pm \sqrt{2}$ **32.** $x = (5 \pm \sqrt{53})/2$ **33.** $x = (-7 \pm \sqrt{33})/2$
34. $x = (-3 \pm \sqrt{17})/2$

Page 53 Quadratic Equations

1. $x = (3 \pm \sqrt{57})/4$ **2.** $x = (-5 \pm \sqrt{65})/4$ **3.** $x = (-6 \pm \sqrt{42})/3$ **4.** $x = (-3 \pm \sqrt{15})/2$
5. $x = (-9 \pm \sqrt{21})/6$ **6.** $x = (4 \pm \sqrt{6})/5$ **7.** $x = (5 \pm \sqrt{7})/6$
8. $x = 5/2$ or $1/2$ **9.** $x = (9 \pm \sqrt{33})/4$
10. $x = (7 \pm \sqrt{109})/6$
11. $x = (-5 \pm 3\sqrt{3})/2$ **12.** $x = (-3 + \sqrt{29})/5$ **13.** $x = -2/3$ or -3 **14.** $x = -1$ or $-5/4$
15. $x = (-7 \pm \sqrt{97})/4$ **16.** $x = -1/2$ or $-4/3$
17. $x = (-5 \pm \sqrt{7})/3$
18. ϕ **19.** $x = (-9 \pm \sqrt{201})/10$ **20.** $x = (-3 \pm 2\sqrt{11})/7$
21. $x = (5 \pm \sqrt{22})/3$ **22.** $x = 4$ or $1/2$ **23.** ϕ
24. $x = (-5 \pm 2\sqrt{15})/7$ **25.** $x = (-11 \pm 65)/4$ **26.** ϕ
27. $x = 1$ or $2/7$ **28.** $x = (5 \pm \sqrt{13})/3$ **29.** ϕ **30.** $x = (3 \pm \sqrt{19})/2$ **31.** $x = (13 \pm \sqrt{209})/4$
32. $x = (-2 \pm \sqrt{19})/5$ **33.** $x = (-1 \pm 2\sqrt{7})/3$
34. $x = 5/3$ or $1/2$

Page 54 Quadratic Equations

1. $x = (-3 \pm \sqrt{65})/4$ 2. $x = (-5 \pm \sqrt{37})/6$ 3. ϕ
4. $x = 1$ or $-3/7$ 5. $x = (2 \pm \sqrt{14})/5$
6. $x = (-7 \pm \sqrt{105})/4$ 7. ϕ 8. $x = (11 \pm \sqrt{445})/18$
9. $x = -1/2$ or -6 10. $x = (-3 \pm \sqrt{2})/7$
11. $x = 1$ or $-5/9$ 12. $x = (-2 \pm 3\sqrt{6})/10$
13. $x = (-7 \pm \sqrt{113})/4$ 14. ϕ
15. $x = (-4 \pm \sqrt{43})/3$ 16. $x = 1$ or $1/5$
17. $x = (11 \pm \sqrt{93})/14$ 18. $x = (3 \pm \sqrt{33})/4$
19. $x = (-5 \pm \sqrt{73})/4$
20. $x = (-7 \pm \sqrt{157})/6$ 21. ϕ 22. $x = 1$ or $1/7$
23. $x = (-9 \pm 3\sqrt{21})/10$ 24. $x = (-11 \pm \sqrt{373})/18$
25. $x = (-7 \pm \sqrt{249})/20$ 26. ϕ 27. $x = (-3 \pm \sqrt{33})/4$
28. $x = -2$ or $-3/5$ 29. $x = (-5 \pm \sqrt{55})/3$ 30. $x = 1$ or $1/10$ 31. $x = (-7 \pm \sqrt{73})/4$ 32. ϕ 33. $x = -1$ or $-7/2$ 34. $x = (5 \pm 2\sqrt{15})/5$

Page 55 Quadratic Equations

1. $x = -2 \pm 2\sqrt{3}$ 2. $x = 3 \pm \sqrt{5}$ 3. $x = 8 \pm 3\sqrt{5}$
4. $x = 3 \pm \sqrt{3}$ 5. $x = -5 \pm 4\sqrt{2}$ 6. $x = 3 \pm 2\sqrt{3}$ 7. ϕ
8. $x = 4 \pm 3\sqrt{7}$ 9. $x = 4$ or 3 10. $x = -3 \pm 5\sqrt{2}$
11. $x = 1 \pm 3\sqrt{2}$ 12. ϕ 13. $x = -6 \pm 2\sqrt{10}$ 14. $x = -8 \pm 3\sqrt{11}$ 15. $x = -5 \pm 2\sqrt{7}$ 16. $x = 6$ or 2 17. ϕ
18. $x = -1 \pm \sqrt{2}$ 19. $x = -3 \pm 6\sqrt{7}$ 20. $x = -2 \pm \sqrt{3}$
21. $x = -5$ or -1 22. ϕ 23. $x = -3 \pm 3\sqrt{10}$ 24. $x = -4 \pm 3\sqrt{2}$ 25. $x = -8 \pm \sqrt{186}$ 26. $x = -4 \pm \sqrt{5}$
27. $x = 8 \pm 4\sqrt{10}$ 28. $x = 3$ or -2 29. $x = 5 \pm 2\sqrt{6}$
30. $x = 3 \pm 2\sqrt{5}$ 31. $x = -3 \pm 4\sqrt{3}$ 32. $x = -5 \pm 4\sqrt{2}$
33. $x = 5 \pm \sqrt{3}$ 34. $x = 10 \pm 6\sqrt{3}$

Page 56 Quadratic Equations

1. $x = 4 \pm 5\sqrt{2}$ 2. $x = -7 \pm 5\sqrt{2}$ 3. $x = 5 \pm 2\sqrt{3}$
4. $x = 4 \pm 3\sqrt{11}$ 5. $x = -7 \pm 2\sqrt{2}$ 6. $x = -3 \pm \sqrt{39}$
7. $x = 5 \pm 2\sqrt{6}$ 8. $x = -3 \pm \sqrt{13}$ 9. $x = 2$ or -7
10. $x = 11 \pm 4\sqrt{7}$ 11. ϕ 12. $x = -5 \pm 3\sqrt{5}$ 13. $x = 2 \pm 3\sqrt{7}$ 14. $x = 5 \pm 5\sqrt{2}$ 15. $x = 7 \pm 2\sqrt{7}$ 16. $x = -3 \pm 2\sqrt{13}$ 17. $x = 8$ or -3 18. ϕ 19. $x = -5 \pm 3\sqrt{7}$
20. $x = 4 \pm 2\sqrt{5}$ 21. ϕ 22. $x = 2$ or -10 23. $x = 7 \pm 3\sqrt{2}$ 24. $x = 8 \pm 3\sqrt{10}$ 25. $x = -3 \pm 2\sqrt{6}$ 26. ϕ
27. $x = -9 \pm 3\sqrt{7}$ 28. $x = -7 \pm 4\sqrt{5}$ 29. $x = 8 \pm 7\sqrt{3}$
30. $x = 3 \pm 2\sqrt{11}$ 31. $x = 2 \pm 4\sqrt{5}$ 32. $x = 2 \pm \sqrt{5}$
33. $x = -3 \pm 2\sqrt{7}$ 34. $x = 6 \pm \sqrt{6}$

Page 57 Quadratic Equations

1. $x = (-3 \pm \sqrt{2})/2$ 2. $x = (1 \pm 6\sqrt{2})/5$ 3. $x = (4 \pm \sqrt{5})/2$ 4. ϕ 5. $x = (5 \pm \sqrt{3})/2$ 6. $x = (6 \pm 5\sqrt{5})/2$
7. $x = (1 \pm 4\sqrt{5})/3$ 8. $x = 3/4$ or $2/5$
9. $x = (-7 \pm \sqrt{145})/6$ 10. $x = (3 \pm \sqrt{149})/14$
11. $x = (7 \pm 2\sqrt{3})/2$ 12. $x = (-1 \pm 5\sqrt{3})/3$
13. $x = (-4 \pm 5\sqrt{5})/2$ 14. $x = (-3 \pm \sqrt{3})/3$ 15. ϕ
16. ϕ 17. ϕ 18. $x = (1 \pm 2\sqrt{5})/3$
19. $x = (3 \pm \sqrt{69})/10$ 20. $x = 5/6$ or $2/3$
21. $x = (-2 \pm 3\sqrt{3})/2$ 22. $x = (19 \pm 4\sqrt{13})/9$ 23. $x = 1/3$ or $1/5$ 24. $x = (7 \pm \sqrt{97})/12$ 25. $x = (-1 \pm \sqrt{6})/2$
26. $x = (7 \pm 3\sqrt{3})/2$ 27. ϕ 28. $x = (3 \pm 2\sqrt{5})/2$
29. $x = (7 \pm 3\sqrt{2})2$ 30. $x = (2 \pm 3\sqrt{3})/2$
31. $x = (7 \pm \sqrt{157})/6$ 32. ϕ 33. $x = (-1 \pm 3\sqrt{2})/2$
34. $x = (1 \pm 2\sqrt{3})/2$

Page 58 Quadratic Equations

1. $x = (1 \pm 2\sqrt{5})/2$ 2. $x = (3 \pm \sqrt{3})/2$ 3. $x = (5 \pm 5\sqrt{2})/2$ 4. $x = (4 \pm 10\sqrt{7})/18$ 5. $x = (-7 \pm 3\sqrt{17})/4$ 6. $x = 3/5$ or $-1/2$ 7. $x = (6 \pm \sqrt{38})/3$ 8. ϕ 9. $x = 2/3$ or $-3/2$ 10. $x = (4 \pm \sqrt{2})/2$

Page 59 The Discriminant

1. 9, two different rational roots   2. 400, two different rational roots   3. 48, two different irrational roots
4. -23, no real roots   5. 768, two different irrational roots
6. 0, one double rational root   7. 320, two different irrational roots   8. 841, two different rational roots
9. -11, no real roots   10. 625, two different rational roots
11. 0, one double rational root   12. 0, one double rational root   13. 108, two different irrational roots   14. 448, two different irrational roots   15. 220, two different irrational roots   16. -832, no real roots

Page 60 The Discriminant

1. 289, two different rational roots   2. 841, two different rational roots   3. 12, two different irrational roots
4. -20, no real roots   5. 0, one double rational root
6. 72, two different irrational roots   7. 768, two different irrational roots   8. 0, one double rational root   9. 192, two different irrational roots   10. 169, two different rational roots   11. 108, two different irrational roots   12. -7, no real roots   13. 972, two different irrational roots   14. 361, two different rational roots   15. 0, one double rational root   16. -96, no real roots

Page 61 Quadratic Equations

1. $x = 3 \pm 2\sqrt{10}$ 2. $x = -1 \pm \sqrt{5}$ 3. $x = 3 \pm \sqrt{5}$
4. $x = 6$ or 2 5. $x = 5$ 6. $x = 2 \pm \sqrt{3}$ 7. $x = 10 \pm 2\sqrt{2}$
8. $x = (-9 \pm \sqrt{69})/2$ 9. $x = 6$ or -11 10. $x = -6 \pm \sqrt{5}$ 11. ϕ 12. $x = -3$ 13. $x = 5 \pm 3\sqrt{7}$ 14. $x = 5$ or 1 15. $x = 13 \pm 5\sqrt{3}$ 16. $x = (3 \pm \sqrt{5})/2$ 17. $x = 4$ or -7 18. $x = 6 \pm 5\sqrt{2}$ 19. $x = (-1 \pm 5\sqrt{3})/2$
20. $x = (-5 \pm 4\sqrt{3})/3$ 21. $x = (17 \pm \sqrt{65})/28$ 22. $x = 1$ or $-2/7$ 23. ϕ 24. $x = (-7 \pm \sqrt{229})/18$ 25. $x = 1$ or $-2/5$ 26. $x = (5 \pm 4\sqrt{2})/2$ 27. $x = (3 \pm 5\sqrt{3})/2$
28. $x = 7/3$ 29. $x = (2 \pm \sqrt{39})/5$ 30. ϕ 31. $x = -3/5$ 32. $x = 2/3$ or $1/4$ 33. $x = (-4 \pm \sqrt{55})/3$
34. $x = (5 \pm \sqrt{613})/14$

Page 62 Quadratic Equations

1. $x = (1 \pm 5\sqrt{5})/2$ 2. $x = 1 \pm 4\sqrt{2}$ 3. $x = 10 \pm 5\sqrt{3}$
4. $x = -5 \pm 3\sqrt{6}$ 5. $x = 5$ or 4 6. $x = (3 \pm 5\sqrt{5})/2$
7. ϕ 8. $x = 13$ 9. $x = 1 \pm 2\sqrt{7}$ 10. $x = (1 \pm 3\sqrt{5})/2$
11. $x = 3 \pm 3\sqrt{2}$ 12. $x = (-5 \pm 5\sqrt{5})/2$ 13. $x = -6$
14. $x = 7$ or -2 15. $x = 2 \pm 3\sqrt{3}$ 16. ϕ 17. $x = 4 \pm 3\sqrt{5}$ 18. $x = -3 \pm 2\sqrt{3}$ 19. $x = (6 \pm 5\sqrt{5})/2$ 20. $x = (4 \pm 3\sqrt{3})/2$ 21. $x = (5 \pm 2\sqrt{3})/3$ 22. $x = (5 \pm \sqrt{265})/12$
23. $x = 3/4$ 24. ϕ 25. $x = (-1 \pm \sqrt{3})/2$
26. $x = (2 \pm 2\sqrt{3})/3$ 27. $x = 11/2$ or 3
28. $x = (4 \pm \sqrt{26})/5$ 29. ϕ 30. $x = (-1 \pm 3\sqrt{2})/2$
31. $x = (5 \pm 5\sqrt{3})/2$ 32. $x = -2/3$
33. $x = (3 \pm 4\sqrt{3})/3$ 34. $x = 11/5$ or 1

Page 63   Word Problems

1. $w = 7$ cm, $l = 12$ cm   2. $w = 30$ cm, $l = 63$ cm   3. $a = 28$ m   4. $b = 36$ cm, $a = 45$ cm   5. $d = 5$ cm   6. $w = 36$ cm, $l = 48$ cm   7. 34 cm × 34 cm × 8 cm   8. Darrell, 7 h; Ginny, 5 h

Page 64   Word Problems

1. $w = 8$ cm, $l = 15$ cm   2. $w = 8$ m, $l = 29$ m   3. $a = 22$ cm   4. $b = 25$ cm, $a = 32$ cm   5. 4 cm   6. 66 cm × 72 cm   7. 38 cm × 38 cm × 5 cm   8. Bob, 5 h; Vera, 1 h

Page 65   Word Problems

1. $d = 20$ cm   2. $l = 29$ cm   3. $a = 4\sqrt{3}$ units   4. $b = 29$ m, $a = 27$ m   5. 30 cm × 24 cm   6. 10 m × 15 m   7. 324 cm²   8. Michael, 13 h; Ty, 7 h

Page 66   Word Problems

1. 14 m   2. $l = 33$ cm   3. $a = 6\sqrt{3}$ units   4. $a = 15$ m, $b = 19$ m   5. 16 cm × 18 cm   6. 24 m × 30 m   7. 900 cm²   8. Bill, 7 h; Bob, 8 h

Page 67   Complex Numbers

1. i 2. $2i$ 3. $4i$ 4. $5i$ 5. $5i\sqrt{2}$ 6. $6i\sqrt{2}$ 7. $4i\sqrt{5}$ 8. $7i\sqrt{2}$ 9. $6i\sqrt{3}$ 10. $5i\sqrt{5}$ 11. $-12 + 6i$ 12. $-14 + 6i$ 13. $5 + 15i$ 14. $12 + 8i$ 15. $5 + 27i$ 16. $2 + 16i$ 17. $13 + i$ 18. $17 + 11i$ 19. 50   20. 18   21. $(5 - 5i)/2$ 22. $2 + 2i$ 23. $(-18 + 12i)/13$ 24. $(4 + 3i)/5$ 25. i 26. $(-5 + 12i)/13$ 27. $x = \pm i$ 28. $x = \pm 2i$ 29. $x = (-3 \pm i\sqrt{7})/2$ 30. $x = (-5 \pm i\sqrt{7})/2$ 31. $x = (-1 \pm i\sqrt{2})/3$ 32. $x = (-3 \pm i\sqrt{31})/4$

Page 68   Complex Numbers

1. $3i$ 2. $6i$ 3. $9i$ 4. $7i$ 5. $2i\sqrt{5}$ 6. $3i\sqrt{3}$ 7. $4i\sqrt{3}$ 8. $4i\sqrt{6}$ 9. $15i$ 10. $5i\sqrt{6}$ 11. $-28 + 24i$ 12. $-30 + 55i$ 13. $-30 + 120i$ 14. $18 + 135i$ 15. $26 + 80i$ 16. $-12 + 164i$ 17. $75 - 25i$ 18. $92 + 28i$ 19. 128   20. 288   21. $(12 - 6i)/5$ 22. $(35 + 7i)/26$ 23. $(3 + 9i)/5$ 24. $(-15 + 25i)/17$ 25. $(5 + 12i)/13$ 26. $(-12 - 35i)/37$ 27. $x = \pm 4i$ 28. $x = \pm 5i$ 29. $x = -1 \pm i\sqrt{6}$ 30. $x = (-5 \pm i\sqrt{19})/2$ 31. $x = (-6 \pm i\sqrt{34})/5$ 32. $x = (-4 \pm i\sqrt{14})/6$

Page 69   Complex Numbers

1. $8i$ 2. $10i$ 3. $11i$ 4. $14i$ 5. $5i\sqrt{7}$ 6. $3i\sqrt{7}$ 7. $3i\sqrt{10}$ 8. $7i\sqrt{3}$ 9. $12i$ 10. $6i\sqrt{5}$ 11. $210 + 180i$ 12. $168 + 154i$ 13. $-360 + 200i$ 14. $-375 + 125i$ 15. $100 + 150i$ 16. $-5i + 340i$ 17. $215 + 30i$ 18. $156 + 80i$ 19. 392   20. 512   21. $(160 - 100i)/89$ 22. $(175 - 75i)/58$ 23. $(-154 + 84i)/157$ 24. $(-104 + 40i)/97$ 25. $(35 + 12i)/37$ 26. $(63 + 16i)/65$ 27. $x = \pm 6i$ 28. $x = \pm 7i$ 29. $x = (-1 \pm i\sqrt{43})/2$ 30. $x = (-1 \pm i\sqrt{35})/2$ 31. $x = (-3 \pm i\sqrt{19})/7$ 32. $x = (-5 \pm 3i\sqrt{15})/16$

Page 70   Complex Numbers

1. $13i$ 2. $15i$ 3. $16i$ 4. $20i$ 5. $10i\sqrt{2}$ 6. $8i\sqrt{2}$ 7. $8i\sqrt{3}$ 8. $7i\sqrt{5}$ 9. $2i\sqrt{13}$ 10. $10i\sqrt{6}$ 11. $210 + 336i$ 12. $-460 + 437i$ 13. $-780 + 364i$ 14. $360 + 480i$ 15. $72 + 209i$ 16. $100 + 246i$ 17. $432 + 174i$ 18. $660 + 405i$ 19. 800   20. 1250   21. $(9 + 9i)/2$ 22. $1 + i$ 23. $(-16 + 40i)/29$ 24. $(28 + 21i)/25$ 25. i 26. $-i$ 27. $x = \pm 8i$ 28. $x = \pm 9i$ 29. $x = (5 \pm i\sqrt{7})/2$ 30. $x = -3 \pm i\sqrt{3}$ 31. $x = (-3 \pm i\sqrt{3})/4$ 32. $x = (-1 \pm i\sqrt{7})/8$

Page 71   Radical Equations

1. ϕ 2. ϕ 3. $x = -5$ 4. $x = -5$ 5. $x = 2$ 6. $x = 2$ 7. ϕ 8. $x = -14$ 9. $x = -6$ or 1   10. $x = -10$ or -1 11. $x = 3$ 12. $x = 2$ 13. $x = 9$ 14. $x = 8$ 15. $x = -1$ 16. $x = -11$ or -10 17. $x = 16$ or 8   18. $x = 6$ or 13   19. $x = -15$ 20. ϕ 21. $x = -7$ 22. $x = -7$ 23. ϕ 24. ϕ 25. $x = -3$ 26. $x = 5$ 27. ϕ 28. ϕ 29. $x = -9$ 30. $x = -2$ 31. $x = 2$ 32. $x = -8$ 33. $x = 97/16$ 34. ϕ

Page 72   Radical Equations

1. $x = -5$ 2. $x = -8$ 3. $x = 5$ 4. ϕ 5. $x = -2$ 6. $x = -6$ 7. $x = -4$ or -3 8. $x = -2$ 9. $x = -3$ 10. $x = -3$ 11. $x = 7$ or 8   12. $x = 4$ or 13   13. $x = 6$ or 3   14. $x = 6$ or 2   15. $x = -11$ 16. $x = -15$ 17. $x = -10$ 18. $x = -10$ 19. $x = -2$ 20. ϕ 21. $x = 12$ 22. $x = 0$ 23. $x = 22$ 24. ϕ 25. $x = -1$ 26. $x = 2$ 27. ϕ 28. ϕ 29. $x = -4$ 30. $x = 0$ 31. $x = -1$ 32. $x = 1$ 33. ϕ 34. $x = 27$

Page 73   Higher Order Radicals

1. 2   2. 3   3. $3\sqrt[3]{2}$ 4. $2\sqrt[3]{2}$ 5. -7 6. -5 7. 2   8. 3   9. $2\sqrt[4]{4}$ 10. $5\sqrt[4]{25}$ 11. 2   12. 5   13. -3 14. $-2\sqrt[5]{8}$ 15. $5x^3y\sqrt[3]{xy}$ 16. $3a^2b^3\sqrt[3]{a}$ 17. $2pq\sqrt[4]{pq^3}$ 18. $4r^2\sqrt[4]{r^2t^2}$ 19. $x = -3$ 20. $x = -2$ 21. $x = 2\sqrt[3]{4}$ 22. $x = -6\sqrt[3]{6}$ 23. $x = \pm 3; \pm 3i$ 24. $x = 7$ 25. $x = \pm 2; \pm 2i$ 26. $x = 6$ 27. $x = 2$ 28. $x = 4$ 29. $x = \pm i$ or $\pm i\sqrt{2}$ 30. $x = \pm i\sqrt{3}$ or $\pm 2i$ 31. $x = \pm\sqrt{6}$ or $\pm i$ 32. $x = \pm 2i\sqrt{2}$ or $\pm\sqrt{6}$

Page 74   Higher Order Radicals

1. 8   2. 5   3. $3\sqrt[3]{3}$ 4. $6\sqrt[3]{6}$ 5. -4 6. -7 7. 5   8. 7   9. $3\sqrt[4]{3}$ 10. $6\sqrt[4]{6}$ 11. 4   12. 3   13. $-3\sqrt[5]{3}$ 14. $-2\sqrt[5]{2}$ 15. $4x^2y^3\sqrt[3]{xy}$ 16. $2a^4b^3\sqrt[3]{b^2}$ 17. $3m^3n^2\sqrt[4]{m^3n^2}$ 18. $7s^3t\sqrt[4]{t^3}$ 19. $x = -4$ 20. $x = -7$ 21. $x = -9$ 22. $x = -6\sqrt[3]{36}$ 23. $x = 4$ 24. $x = 8$ 25. $x = 5$ 26. $x = 9$ 27. $x = 3$ 28. $x = 6$ 29. $x = \pm i\sqrt{2}$ or $2i$ 30. $x = \pm 3i$ or $\pm i\sqrt{3}$ 31. $x = \pm\sqrt{7}$ or $\pm 2i$ 32. $x = \pm\sqrt{2}$ or $\pm i\sqrt{7}$

Page 75   Introduction to Functions

1. yes   2. yes   3. yes   4. no   5. yes   6. yes   7. yes   8. yes   9. no   10. no   11. no   12. yes   13. yes   14. yes   15. yes   16. yes   17. yes   18. no   19. yes   20. yes   21. yes   22. no

Page 76   Introduction to Functions

1. yes   2. no   3. yes   4. yes   5. yes   6. no   7. yes   8. no   9. yes   10. yes   11. yes   12. yes   13. yes   14. yes   15. yes   16. yes   17. no   18. yes   19. yes   20. no   21. yes   22. no

Page 77   Functional Notation

1. 9   2. 15   3. -1 4. -13 5. 23   6. 29   7. -26 8. -31 9. 6   10. -30 11. 2275   12. 697   13. 7   14. 7   15. 15   16. 15   17. 13   18. 13   19. 9   20. 9   21. 16   22. 16   23. 13   24. 13   25. 5   26. 5   27. 8   28. 9

Page 78   Functional Notation

1. -22 2. -38 3. -6 4. 54   5. -34 6. 22   7. 18   8. -4 9. 8   10. -22 11. 51   12. -247 13. -4 14. -4 15. 1   16. 1   17. -2 18. -2 19. -3 20. 2   21. -1 22. 1   23. 6   24. -1 25. 9   26. 3   27. 1   28. -1

Page 79 Graphing Functions

1. straight line passing through $(0, 3)$ and $(1, 1)$
2. straight line passing through $(0, -1)$ and $(3, 1)$
3. V-shaped graph opening upward; vertex at $(0, 0)$; one ray passes through $(3, 3)$; the other ray passes through $(-3, 3)$
4. V-shaped graph opening upward; vertex at $(0, 2)$; one ray passes through $(2, 4)$; the other ray passes through $(-2, 4)$
5. V-shaped graph opening upward; vertex at $(0, 0)$; one ray passes through $(2, 4)$; the other ray passes through $(-2, 4)$
6. V-shaped graph opening upward; vertex at $(0, -1)$; one ray passes through $(1, 2)$; the other ray passes through $(-1, 2)$

Page 80 Graphing Functions

1. straight line passing through $(0, -2)$ and $(1, 2)$
2. straight line passing through $(0, -1)$ and $(4, 2)$
3. V-shaped graph opening downward; vertex at $(0, 0)$; one ray passes through $(3, -3)$; the other ray passes through $(-3, -3)$ 4. V-shaped graph opening downward; vertex at $(0, 3)$; one ray passes through $(3, 0)$; the other ray passes through $(-3, 0)$ 5. V-shaped graph opening downward; vertex at $(0, 0)$; one ray passes through $(1, -3)$; the other ray passes through $(-1, -3)$ 6. V-shaped graph opening downward; vertex at $(0, -1)$; one ray passes through $(1, -3)$; the other ray passes through $(-1, -3)$

Page 81 Graphing Quadratic Functions

Parabola opening upward, symmetric about the y-axis, with vertex at $(0, 0)$.

x	-4	-3	-2	-1	0	1	2	3	4
y	16	9	4	1	0	1	4	9	16

1. $(0, 0)$ 2. minimum 3. $x = 0$ 4. $(0, 0)$

Page 82 Graphing Quadratic Functions

Parabola opening downward, symmetric about the y-axis, with vertex at $(0, 0)$.

x	-4	-3	-2	-1	0	1	2	3	4
y	-16	-9	-4	-1	0	-1	-4	-9	-16

1. $(0, 0)$ 2. maximum 3. $x = 0$ 4. $(0, 0)$

Page 83 Graphing Quadratic Functions

Parabola opening upward, symmetric about the y-axis, with vertex at $(0, 0)$. Graph is narrower than graph of $y = x^2$.

x	-4	-3	-2	-1	0	1	2	3	4
y	32	18	8	2	0	2	8	18	32

1. $(0, 0)$ 2. minimum 3. $x = 0$ 4. $(0, 0)$ 5. narrower

Page 84 Graphing Quadratic Functions

Parabola opening downward, symmetric about the y-axis, with vertex at $(0, 0)$. Graph is narrower than graph of $y = x^2$.

x	-4	-3	-2	-1	0	1	2	3	4
y	-48	-27	-12	-3	0	-3	-12	-27	-48

1. $(0, 0)$ 2. maximum 3. $x = 0$ 4. $(0, 0)$
5. narrower

Page 85 Graphing Quadratic Functions

1. Parabola opening upward, symmetric about the y-axis with vertex at $(0, -1)$. Graph has same shape as $y = x^2$.
2. Parabola opening upward, symmetric about the y-axis, with vertex at $(0, 2)$. Graph has same shape as $y = x^2$.
3. Parabola opening downward, symmetric about the y-axis, with vertex at $(0, 3)$. Graph has same shape as $y = x^2$.
4. Parabola opening downward, symmetric about the y-axis, with vertex at $(0, -1)$. Graph has same shape as $y = x^2$.
5. Parabola opening upward, symmetric about the y-axis, with vertex at $(0, 1)$. Graph is narrower than graph of $y = x^2$. 6. Parabola opening downward, symmetric about the y-axis, with vertex at $(0, 2)$. Graph is wider than graph of $y = x^2$.

Page 86 Graphing Quadratic Functions

1. Parabola opening upward, symmetric about $x = 1$, with vertex at $(1, 0)$. Graph has same shape as $y = x^2$.
2. Parabola opening upward, symmetric about $x = -2$, with vertex at $(-2, 0)$. Graph has same shape as $y = x^2$.
3. Parabola opening upward, symmetric about $x = 3$, with vertex at $(3, 0)$. Graph has same shape as $y = x^2$.
4. Parabola opening upward, symmetric about $x = -1$, with vertex at $(-1, 0)$. Graph has same shape as $y = x^2$.
5. Parabola opening downward, symmetric about $x = 1$, with vertex at $(1, 0)$. Graph is narrower than graph of $y = x^2$.
6. Parabola opening upward, symmetric about $x = -2$, with vertex at $(-2, 0)$. Graph is wider than graph of $y = x^2$.

Page 87 Analysis of Quadratic Functions

1. $(0, 1)$; min; $x = 0$; none; same 2. $(0, 3)$; min; $x = 0$; none; narrower 3. $(0, -2)$; min; $x = 0$; $(2, 0)$, $(-2, 0)$; wider 4. $(0, -3)$; max; $x = 0$; none; narrower 5. $(0, 5)$; min; $x = 0$; none; wider 6. $(-3, 0)$; min; $x = -3$; $(-3, 0)$; same 7. $(2, 0)$; min; $x = 2$; $(2, 0)$; same 8. $(-1, 0)$; max; $x = -1$; $(-1, 0)$; narrower 9. $(4, 0)$; max; $x = 4$; $(4, 0)$; wider 10. $(5, 0)$; min; $x = 5$; $(5, 0)$; narrower

Page 88 Analysis of Quadratic Functions

1. $(0, -5)$; min; $x = 0$; $(\pm \sqrt{5}, 0)$; same 2. $(0, 3)$; min; $x = 0$; none; same 3. $(0, -1)$; max; $x = 0$; none; narrower 4. $(0, 5)$; min; $x = 0$, none; wider 5. $(0, 2)$; max; $x = 0$; $(\pm \sqrt{2}/2, 0)$; narrower 6. $(2, 0)$; min; $x = 2$; $(2, 0)$; same 7. $(-5, 0)$; min; $x = -5$; $(-5, 0)$; same 8. $(4, 0)$; min; $x = 4$; $(4, 0)$; narrower 9. $(-2, 0)$; min; $x = -2$; $(-2, 0)$; wider 10. $(1, 0)$; min; $x = 1$; $(1, 0)$; narrower

Page 89 Graphing Quadratic Functions

1. Parabola opening upward, symmetric about $x = 1$, with vertex at $(1, 2)$. Graph has same shape as $y = x^2$.
2. Parabola opening upward, symmetric about $x = -1$, with vertex at $(-1, -1)$. Graph has same shape as $y = x^2$.
3. Parabola opening downward, symmetric about y-axis, with vertex at $(0, -3)$. Graph is narrower than graph of $y = x^2$. 4. Parabola opening downward, symmetric about $x = 2$, with vertex at $(2, 2)$. Graph is wider than graph of $y = x^2$.
5. Parabola opening upward, symmetric about $x = -3$, with vertex at $(-3, -4)$. Graph is wider than graph of $y = x^2$.
6. Parabola opening upward, symmetric about $x = 1$, with vertex at $(1, 1)$. Graph is narrower than graph of $y = x^2$.

Page 90 Graphing Quadratic Functions

1. Parabola opening upward, symmetric about $x = 2$, with vertex at $(2, 1)$. Graph has same shape as $y = x^2$.
2. Parabola opening downward, symmetric about $x = -2$, with vertex at $(-2, -1)$. Graph has same shape as $y = x^2$.
3. Parabola opening upward, symmetric about $x = 4$, with vertex at $(4, -3)$. Graph is wider than graph of $y = x^2$.
4. Parabola opening upward, symmetric about $x = 1$, with vertex at $(1, -1)$. Graph is narrower than graph of $y = x^2$.
5. Parabola opening downward, symmetric about $x = -2$, with vertex at $(-2, 1)$. Graph is narrower than graph of $y = x^2$.
6. Parabola opening downward, symmetric about $x = 2$, with vertex at $(2, 2)$. Graph is narrower than graph of $y = x^2$.

Page 91 Analysis of Quadratic Functions

1. $(-3, -4)$; min; $x = -3$; $(-5, 0)$, $(-1, 0)$; same
2. $(1, 2)$; min; $x = 1$; $(3, 0)$, $(1, 0)$; same **3.** $(4, 3)$; min; $x = 4$; none; narrower **4.** $(-1, -4)$; max; $x = 1$; none; narrower
5. $(2, 4)$; min; $x = 2$; none; wider **6.** $(4, 1)$; max; $x = 4$; $([20 \pm \sqrt{15}]/5, 0)$; narrower **7.** $(1, -6)$; min; $x = 1$; $([2 \pm \sqrt{6}]/2, 0)$; narrower **8.** $(-1, -5)$; min; $x = -1$; $([-3 \pm 2\sqrt{30}]/3, 0)$; wider **9.** $(-3, -4)$; max; $x = -3$; none; wider **10.** $(1, 1)$; max; $x = 1$; $([10 \pm \sqrt{10}]/10, 0)$; narrower

Page 92 Analysis of Quadratic Functions

1. $(2, 3)$; min; $x = 2$; none; same **2.** $(-5, -1)$; min; $x = -5$; $(-4, 0)$, $(-6, 0)$; same **3.** $(-6, -2)$; max; $x = -6$; none; narrower **4.** $(3, 1)$; min; $x = 3$; none; narrower **5.** $(5, 2)$; max; $x = 5$; $([15 \pm 2\sqrt{6}]/3, 0)$; wider **6.** $(5, 3)$; min; $x = 5$; none; narrower **7.** $(7, 2)$; max; $x = 7$; $([14 \pm \sqrt{10}]/2, 0)$; wider **8.** $(-3, -6)$; max; $x = -3$; none; narrower **9.** $(3, 7)$; min; $x = 3$; none; narrower **10.** $(2, 9)$; min; $x = 2$; none; narrower

Page 93 Analysis of Quadratic Functions

1. $(-1, 1)$; $y = (x - 0)^2 + 0$ **2.** $(-1, -1)$; $y = -(x - 0)^2 + 0$ **3.** $(-4, 2)$; $y = (1/8)(x - 0)^2 + 0$ **4.** $(-3, 10)$; $y = (10/9)(x - 0)^2 + 0$ **5.** $(3, 8)$; $y = (x - 0)^2 - 1$ **6.** $(-4, 10)$; $y = (5/16)(x - 0)^2 + 5$ **7.** $(-1, 7)$; $y = 5(x - 0)^2 + 2$ **8.** $(2, 5)$; $y = 2(x - 0)^2 - 3$ **9.** $(3, 7)$; $y = (7/4)(x - 5)^2 + 0$ **10.** $(-4, 10)$; $y = (10/9)(x + 1)^2 + 0$ **11.** $(1, 2)$; $y = (1/2)(x - 3)^2 + 0$ **12.** $(-3, 6)$; $y = 6(x + 2)^2 + 0$ **13.** $(0, 5)$; $y = (1/2)(x - 2)^2 + 3$ **14.** $(-1, 10)$; $y = 3(x - 1)^2 - 2$

Page 94 Analysis of Quadratic Equations

1. $(-2, 3)$; $y = (3/4)(x - 0)^2 + 0$ **2.** $(-3, 1)$; $y = (1/9)(x - 0)^2 + 0$ **3.** $(-1, -7)$; $y = -7(x - 0)^2 + 0$ **4.** $(-3, -27)$; $y = -3(x - 0)^2 + 0$ **5.** $(-2, 8)$; $y = (x - 0)^2 + 4$ **6.** $(-3, -20)$; $y = -2(x - 0)^2 - 2$ **7.** $(-5, 16)$; $y = (3/5)(x - 0)^2 + 1$ **8.** $(-2, -15)$; $y = (-9/4)(x - 0)^2 - 6$ **9.** $(-6, 4)$; $y = (x + 4)^2 + 0$ **10.** $(3, 3)$; $y = 3(x - 2)^2 + 0$ **11.** $(-8, 3)$; $y = (1/3)(x + 5)^2 + 0$ **12.** $(15, -20)$; $y = (-4/5)(x - 10)^2 + 0$ **13.** $(0, -11)$; $y = (-2/3)(x - 3)^2 - 5$ **14.** $(-3, 30)$; $y = 7(x + 1)^2 + 2$

Page 95 Zeros of Quadratic Functions

1. $x = 1 \pm \sqrt{5}$ **2.** $x = -3 \pm \sqrt{2}$ **3.** $x = -1$ or -3
4. $x = 4 \pm \sqrt{3}$ **5.** $x = 4$ or 8 **6.** $x = -5 \pm \sqrt{2}$ **7.** ϕ
8. ϕ **9.** $x = (8 \pm \sqrt{6})/2$ **10.** $x = -3 \pm 2\sqrt{2}$ **11.** $x = 4$ or -2 **12.** $x = (10 \pm \sqrt{5})/5$ **13.** $x = (-6 \pm \sqrt{21})/3$
14. ϕ **15.** ϕ **16.** $x = (3 \pm 2\sqrt{15})/3$
17. $x = (12 \pm \sqrt{30})/6$ **18.** $x = (-2 \pm \sqrt{6})/2$ **19.** ϕ
20. ϕ **21.** $x = (2 \pm \sqrt{10})/2$ **22.** $x = 3$ or 1 **23.** ϕ
24. ϕ **25.** $x = (2 \pm \sqrt{3})/2$ **26.** $x = (8 \pm \sqrt{2})/2$
27. $x = (-10 \pm \sqrt{30})/5$ **28.** $x = (28 \pm \sqrt{21})/7$
29. $x = 3 \pm \sqrt{14}$ **30.** $x = (-4 \pm \sqrt{5})/2$ **31.** ϕ
32. $x = -1 \pm \sqrt{2}$ **33.** $x = -1 \pm \sqrt{5}$ **34.** ϕ

Page 96 Zeros of Quadratic Functions

1. $x = 2 \pm \sqrt{3}$ **2.** $x = -3$ or -5 **3.** $x = -6 \pm \sqrt{5}$
4. $x = 5 \pm \sqrt{6}$ **5.** $x = -1 \pm \sqrt{3}$ **6.** $x = 4 \pm \sqrt{2}$ **7.** $x = -5$ or -1 **8.** $x = -6 \pm \sqrt{3}$ **9.** $x = (9 \pm \sqrt{3})/3$ **10.** $x = (-6 \pm \sqrt{10})/2$ **11.** ϕ **12.** ϕ **13.** $x = (-2 \pm \sqrt{2})/2$
14. $x = -1 \pm \sqrt{2}$ **15.** $x = 2 \pm 2\sqrt{2}$ **16.** ϕ
17. $(-8 \pm \sqrt{30})/2$ **18.** $x = (5 \pm \sqrt{30})/5$ **19.** ϕ **20.** ϕ
21. $x = (21 \pm \sqrt{14})/7$ **22.** $x = (3 \pm 2\sqrt{3})/3$ **23.** ϕ
24. $x = (-4 \pm \sqrt{14})/2$ **25.** $x = (3 \pm \sqrt{3})/3$ **26.** $x = (9 \pm 2\sqrt{3})/3$ **27.** $x = (-15 \pm \sqrt{35})/5$ **28.** ϕ **29.** ϕ
30. $x = (5 \pm \sqrt{30})/5$ **31.** $x = (20 \pm \sqrt{30})/10$ **32.** $x = (-6 \pm \sqrt{15})/3$ **33.** $x = (2 \pm \sqrt{10})/2$
34. $x = (-6 \pm 2\sqrt{3})/3$

Page 97 Completing the Square

1. $(5, 2)$ **2.** $(-3, -5)$ **3.** $(-1, -7)$ **4.** $(4, 3)$ **5.** $(2, -6)$
6. $(-10, -5)$ **7.** $(-7, -9)$ **8.** $(3, 2)$ **9.** $(5, -4)$
10. $(-8, -3)$ **11.** $(3, 6)$ **12.** $(-2, -5)$ **13.** $(1, 1)$
14. $(5, 6)$ **15.** $(-2, -1)$ **16.** $(2, 3)$ **17.** $(3, 4)$
18. $(-1, -2)$ **19.** $(3, 2)$ **20.** $(-7, -3)$ **21.** $(-6, -3)$
22. $(4, 4)$ **23.** $(10, -5)$ **24.** $(5, -4)$ **25.** $(-4, 5)$
26. $(3, 7)$ **27.** $(6, -7)$ **28.** $(8, 7)$ **29.** $(-3, 8)$
30. $(-5, -1)$ **31.** $(-9, 3)$ **32.** $(7, 9)$ **33.** $(5, -6)$
34. $(4, 2)$

Page 98 Completing the Square

1. $(-1, 4)$ **2.** $(2, -1)$ **3.** $(7, -2)$ **4.** $(-5, 3)$ **5.** $(1, 1)$
6. $(3, -2)$ **7.** $(-4, 5)$ **8.** $(5, -6)$ **9.** $(-6, 6)$
10. $(-2, -4)$ **11.** $(4, -3)$ **12.** $(6, 3)$ **13.** $(-2, -3)$
14. $(4, -2)$ **15.** $(1, -1)$ **16.** $(-1, 3)$ **17.** $(2, -4)$
18. $(-4, -8)$ **19.** $(-5, 2)$ **20.** $(2, 7)$ **21.** $(-6, -10)$
22. $(3, -2)$ **23.** $(3, -7)$ **24.** $(-3, -5)$ **25.** $(-4, 2)$
26. $(-6, 5)$ **27.** $(-10, -6)$ **28.** $(5, 5)$ **29.** $(-3, -7)$
30. $(1, 8)$ **31.** $(11, 4)$ **32.** $(-7, 7)$ **33.** $(7, 6)$
34. $(6, 3)$

Page 99 Maximum/Minimum Word Problems

1. 400 ft. **2.** 8 s **3.** 12,769 m^2 **4.** 128 and 128
5. 10 and 10 **6.** 12 cm and 12 cm **7.** 1100 watts
8. 30 stories **9.** 5 **10.** 15 boxes

Page 100 Maximum/Minimum Word Problems

1. 784 ft **2.** 6 s **3.** 22,500 m^2 **4.** 150 and 150 **5.** 15 and 15 **6.** 18 cm and 18 cm **7.** 605 watts **8.** 20 stories
9. 11 **10.** 30 boxes